Information theory, evolution, and the origin of life

Information Theory, Evolution, and the Origin of Life presents a timely introduction to the use of information theory and coding theory in molecular biology. The genetical information system, because it is linear and digital, resembles the algorithmic language of computers. George Gamow pointed out that the application of Shannon's information theory breaks genetics and molecular biology out of the descriptive mode into the quantitative mode, and Dr. Yockey develops this theme, discussing how information theory and coding theory can be applied to molecular biology. He discusses how these tools for measuring the information in the sequences of the genome and the proteome are essential for our complete understanding of the nature and origin of life. The author writes for the computer competent reader who is interested in evolution and the origins of life.

Hubert P. Yockey is a former director of the Pulsed Radiation Facility at the US Army's Aberdeen Proving Ground, Maryland. He is the author of *Information Theory and Molecular Biology* (1992).

Information theory, evolution, and the origin of life

HUBERT P. YOCKEY

CAMBRIDGE UNIVERSITY PRESS
Cambridge, New York, Melbourne, Madrid, Cape Town, Singapore,
São Paulo, Delhi, Dubai, Tokyo, Mexico City

Cambridge University Press
32 Avenue of the Americas, New York, NY 10013-2473, USA

www.cambridge.org
Information on this title: www.cambridge.org/9780521169585

First published 2005
Reprinted 2006
First paperback edition 2010

A catalog record for this publication is available from the British Library

Library of Congress Cataloging in Publication data
Yockey, Hubert P.
Information theory, evolution, and the origin of life / Hubert P. Yockey.
  p. cm.
Includes bibliographical references (p.).
ISBN 0-521-80293-8 (hardback : alk. paper)
1. Molecular biology.   2. Information theory in biology.   3. Evolution (Biology)
4. Life–Origin.   I. Title.
QH506.Y634   2004
572.8 – dc22   2004054518

ISBN 978-0-521-80293-2 Hardback
ISBN 978-0-521-16958-5 Paperback

It must be considered that there is nothing more difficult to carry out nor more doubtful of success, nor more dangerous to handle, than to initiate a new order of things. For the reformer has enemies in all those who profit by the old order, and only lukewarm defenders in all those who would profit by the new order, this lukewarmness arising partly for fear of their adversaries, who have the laws in their favor; and partly from the incredulity of men, who do not truly believe in anything new until they have had actual experience of it.

Niccolò Machiavelli (1469–1519), *The Prince*, Chapter 6.

Contents

Preface   *page* ix

1 The genetic information system   1
2 James Watson, Francis Crick, George Gamow, and
  the genetic code   8
3 The Central Dogma of molecular biology   20
4 The measure of the information content in the
  genetic message   27
5 Communication of information from the genome to the
  proteome   33
6 The information content or complexity of
  protein families   57
7 Evolution of the genetic code and its modern
  characteristics   93
8 Haeckel's *Urschleim* and the role of the Central Dogma
  in the origin of life   114
9 Philosophical approaches to the origin of life   149
10 The error catastrophe and the hypercycles of Eigen and
   Schuster   158
11 Randomness, complexity, the unknowable,
   and the impossible   164
12 Does evolution need an intelligent designer?   176
13 Epilogue   182

Mathematical appendix	191
Glossary	213
References	219
Index	251

program in computer memories and the genetic message recorded in DNA (Chaitin, 1979).

The genetic information system is essentially a digital data recording and processing system. The fundamental axiom in genetics and molecular biology, which justifies the application of Shannon's information and coding theory, is the *sequence hypothesis* and the *digital* rather than the *analog* or *blending* character (Jenkin, 1867) of inheritance as Darwin (1809–82) and his contemporaries believed (Fisher, 1930).

Watson and Crick's solution of the structure of DNA and its application in biology would not have been so important if it had not been for their famously coy remark:

> It has not escaped our notice that the specific pairing we have postulated immediately suggests a copying mechanism for the genetic material. (*Nature* 1953)

A fundamental question in genetics is, how does the cell divide into two cells both containing the same genetics? Here, at one stroke, was the solution nicely framed by reductionism!

I show in this book that only because the genetic message is *segregated, linear, and digital* can it be transmitted from the origin of life to all present organisms and will be transmitted to all future life. This establishes Darwin's theory of evolution as firmly as any in science. The same genetic code, the same DNA, the same amino acids, and the genetic message unite all organisms, independent of morphology.

The genetic message recorded in the DNA of every living organism is unique to that individual. The relationship and evolution among animals and plants can now be determined by comparing DNA sequences rather than relying on morphology. Genetic information is being applied to genomic medical practice and genetic counseling for the benefit of patients. Sickle-cell anemia is a blood disorder that is an important example of the role of DNA in the placement of amino acids in the sequences of amino acids that form hemoglobin. It is so named because the red blood cells that are normally round are shaped like a sickle. Hemoglobin is composed of four chains of amino acids. Fundamental to this disease, at site 6 in the β chain, glutamic acid is replaced by valine. The identification of this genetic disorder in the structure of hemoglobin with the symptoms of sickle anemia was made by Linus Pauling (1901–94) and is one of his more important discoveries (Pauling, 1949).

Preface

This book introduces the general reader and the specialist to the new order of things in evolution, the origin of life on Earth, and the question of life on Mars and Europa and elsewhere in the universe. Although there are many fields of biology that are essentially descriptive, with the application of information theory, theoretical biology can now take its place with theoretical physics without apology. Thus biology has become a quantitative and computational science as George Gamow (1904–68) suggested. By employing information theory, comparisons between the genetics of organisms can now be made quantitatively with the same accuracy that is typical of astronomy, physics, and chemistry.

Spacecraft send messages to Earth as they pass the outer planets – Mars, Jupiter, Saturn, Uranus, Neptune, and Pluto – in spite of the small amount of energy available. Enormous amounts of data and information flow about on the Internet. Huge sums of money are transferred every day. Errors in these communications cannot be tolerated. Claude Shannon (1916–2001) showed that this is accomplished because communication is *segregated, linear, and digital* so that sufficient redundance can be introduced in communication codes to overcome errors. Furthermore, he showed that these signals, which contain messages, can be measured in *bits* and *bytes*, terms that are familiar to computer users.

Watson and Crick discovered that there is a genetic message, recorded in the digital sequence of nucleotides in DNA, that controls the formation of protein and of course all biological processes. The message in the genetic information system is *segregated, linear, and digital* and can be measured in *bits* and *bytes*. Computer users will notice the isomorphism between the

DNA now plays a role in forensics identification that is far more important than fingerprints. Forensics has reached new levels of certainty. A number of guilty people have been convicted, and others, falsely accused by conventional methods, have been vindicated.

This is a monograph and not an encyclopedia so I have not considered it necessary to call attention to papers I believe did not make an important contribution or those which are incorrect. I have included in the references only those I felt contributed to the point I was making. Some readers may think that I have neglected an important paper here and there. I acknowledge that this may be the case, but there are times when one must hew to the line and let the chips fall where they may.

This monograph follows my interest in the subject, which was first attracted by the work of Dr. Henry Quastler (1908–63). With his collaboration I organized the *Symposium on Information Theory in Biology* at Gatlinburg, Tennessee, in October 1956. I am indebted to the late Professor Thomas Hughes Jukes (1906–99), whose strong recommendations resulted in my original papers being published. Many of Professor Jukes' important contributions to molecular biology have shaped the ideas presented in this book, particularly those concerning the evolution of the genetic code. I am grateful to Dr. Gregory J. Chaitin, whose original and seminal work in algorithmic information theory is reflected throughout the book. I appreciate the efforts of Dr. David Abel and Mr. John Tomlinson, who read the manuscript and made important corrections and comments. My daughter, Cynthia Ann Yockey, edited this manuscript from proposal to final draft and contributed much to improve the clarity and organization of the material. My editor at Cambridge University Press, Dr. Katrina Halliday, organized the review of the manuscript and arranged for the publication. I appreciate her patience during the writing. Without the contribution of these people I would not have been able to write this book. The reference material is up to date as of February 20, 2004.

Hubert P. Yockey
Bel Air, Maryland, USA

1

The genetic information system

Socrates: Every sort of confusion like these is to be found in our minds; and it is this weakness in our nature that is exploited, with a quite magical effect, by many tricks of illusion, like scene-painting and conjuring.

Glaucon: *True.*

Socrates: But satisfactory means have been found for dispelling these illusions by measuring, counting and weighing. We are no longer at the mercy of apparent differences of size and quantity and weight; the faculty which has done the counting, measuring or weighing takes control instead. And this can only be the work of the calculating or reasoning element in the soul.

The Republic, Book X, Plato (428–348 B.C.),
translated by Francis M. Cornford, Oxford University Press.

1.1 Expressing knowledge in numbers

Socrates (*The Republic*, Book VI, p.745) had noted, in an earlier conversation with Glaucon, that students of geometry and reckoning first set up postulates appropriate to each branch of science, treating them as known absolute assumptions, taking it for granted that they are obvious to everybody. Thus, as Socrates taught us, the essence of science is measuring, counting, and weighing together with reasoning from postulates or axioms. This breaks molecular biology out of sophisticated *Just So Stories* (Kipling, 1902) into the quantitative mode, used by natural scientists (Gamow, 1954; Wolynes, 1998).

Hermann Rorschach (1884–1922), a Swiss psychiatrist, analyzed the interpretations, by his subjects, of ten standard inkblots to probe their thoughts. There is a danger that we may be looking at the shapes of Rorschach inkblots, so to speak, and seeing what we want to see when science attempts to proceed from qualitative arguments. The more discussions can be made quantitative and avoid *ad hoc* explanations, the better our understanding of biology is served. There are no "other ways of knowing" in science. *The absence of evidence is evidence of absence.*

The laws of physics and chemistry are much like the rules of a game such as football. The referees see to it that these laws are obeyed but that does not predict the winner of the Super Bowl. There is not enough information

1

in the rules of the game to make that prediction. That is why we play the game. Chaitin (1985, 1987a) has examined the information content of the laws of physics by actually programming them. He finds the information content amazingly small.

The reason that there are principles of biology that cannot be derived from the laws of physics and chemistry lies simply in the fact that the genetic information content of the genome for constructing even the simplest organisms is much larger than the information content of these laws (Yockey, 1992).

1.1.1 The definition of life and Louis Pasteur

Mr. Justice Potter Stewart (1915–85, U.S. Supreme Court) said he couldn't define pornography, but he knew it when he saw it. It is often said that no broadly accepted definition of life exists. Like Edgar Allan Poe's (1809–48) *The Purloined Letter*, the definition of life has been in plain sight since 1848. One of Louis Pasteur's (1822–95) more important discoveries, relevant to the nature and origin of life, is that ammonium tartrate tetrahydrate when made from grapes has only the left-handed molecules, Pasteur (1848, 1922). When examined in a polarimeter, they are found to rotate the plane of polarization of light to the left. Ammonium tartrate tetrahydrate made synthetically is racemic, that is, composed of equal numbers of right-handed and left-handed molecules. The human hand is chiral. Each hand is the mirror image of the other. Neither can be superimposed on the other.

Pasteur carefully selected the two kinds of crystals, called optical isomers, and found that each rotated the plane of polarization in opposite directions, one left and the other right. He prepared a synthetic ammonium tartrate tetrahydrate solution and contaminated it with a mold. The solution became more optically active with time. It followed that the mold was using only the left-handed ammonium tartrate molecules. What a delicate appetite that mold had! This achievement of Pasteur is the first demonstration of chiral molecules as an essential and unique element in biology. It can serve as a definition of life, as any substance composed of only one optical isomer must have come from life (Section 8.1.3).

An additional criterion for this book is:

> The existence of a genome and the genetic code divides living organisms from nonliving matter. There is nothing in the physico-chemical world that remotely resembles reactions being determined by a sequence and codes between sequences.

1.1.2 The work of Gregor Mendel (1822–84) leading to molecular biology and genetics

In the nineteenth century, intuition led many to believe that the inheritance of characteristics, such as tall and short, would yield a blending of these traits and produce plants of medium height (Jenkin, 1867). The theory of blending inheritance predicts either the disappearance of favored traits or that mutations must be several thousand times as frequent as they are known to be (Fisher, 1930). Therefore, Jenkin concluded that this was evidence that Darwinian evolution would never occur.

However, the Gregor Mendel's experiments with strains of pea (Mendel, 1865) proved that inheritance is *segregated* and does not blend. This was the first step to the molecular biology and genetics we have today. The structure of DNA found by Watson (1928–) and Crick (1916–2004) could have been just that of another large molecule, such as hemoglobin, if it had not been that DNA carries the genetic message that is transferred to the proteome by the genetic code. Their work completed the modern view that the message in the genetic information system is *segregated, linear, and digital*. Watson and Crick's finding is just as much a new axiom in science, as Max Planck's discovery that Newton's particles of light are electromagnetic wave packets and the frequency, v, is related to the energy, E, by $E = hv$ where h is Planck's constant.

The genetical information system, because it is *segregated, linear, and digital*, resembles the algorithmic language by which a computer completes its logical operation (C. H. Bennett, 1973; Chaitin, 1979). Computer users are well aware that the amount of information in a sequence or a message can be measured without regard for its meaning. A computer user buying a floppy disk or a hard drive does not expect it to hold either more or less information depending on whether it will be used to store children's drawings or translations of the plays of Sophocles. Information theory and coding theory and their tools of measuring the information in the sequences of the genome and the proteome are essential to understanding the crucial questions of the nature and the origin of life.

1.1.3 Evolution and the sequencing of DNA

As all the living forms of life are the lineal descendants of those which lived long before the Cambrian epoch, we may feel certain that the ordinary progression by generation has never once been broken and no cataclysm has devastated the world. " . . . from so

simple a beginning endless forms most beautiful and most won-
derful have been, and are being evolved. (Darwin, 1872, Ch. XV)

The recent accomplishments in the sequencing of the DNA of the human
genome as well as those of a number of other organisms establishes the
remarks of Darwin beyond question. Darwin was concerned with "missing
links" and based much of his "one long argument" (*Origin of Species*, 1872
edition, Ch. XV) on comparative morphology. The old arguments based on
"missing links" proposed in opposition of Darwin's theory are no longer
relevant. For that reason, foolish discussions on either side of the debate on
Darwinism about how the giraffe got his long neck are no longer pertinent.
Although the details may be unknowable, there is indeed a phylogenetic
evolutionary message or signal from which all organisms have branched
(Woese, 1998, 2000, 2002) (see Section 11.2.3). The e-mail one sends to
colleagues traces its way through the Internet from source to destination.
By the same token, the "code-script," as Schrödinger (1992) called it, unites
all living things on Earth.

The transmission of genetic messages for more than 3.85 billion years
since the origin of life (Mojzsis et al. 1999; Woese, 2000), with modification
and diversification by evolution, could have been done *only* because the
message in the genome is *segregated, linear, and digital* (Chapter 12). It
is impossible to remove the effect of noise in analog signals. Early analog
records of the glorious voice of Enrico Caruso (1873–1921) do not compare
with the modern digital recordings of the Three Tenors: Plácido Domingo,
José Carreras, and Luciano Pavarotti. Shannon's Channel Capacity Theorem
(Shannon, 1948) showed how to eliminate the effect of noise as much as we
wish by digitizing the signal. The digital revolution has now provided digital
television eliminating noise almost to the theoretical limit. Even cameras
are now digital.

Evolution would be quite impossible if inheritance were by analog means.
Nevertheless, distinguished biologists Szathmáry and Maynard Smith
(1997) wrote:

> To explain the origin of life, we need to explain the origin of heredity
> in terms of chemistry.

Morowitz et al. (2000) wrote:

> A small number of selection rules generates a very constrained
> subset, suggesting that this is the type of reaction model that will
> prove useful in the study of biogenesis. The model indicates that
> the metabolism shown in the universal chart of pathways may be

central to the origin of life, is emergent from organic chemistry, and may be unique.

2.1 The contributions of Niels Bohr

Niels Bohr (1885–1962) proposed that life is *consistent* with but *undecidable* or *unknowable* by human reasoning from physics and chemistry. Bohr (1933) made this point in his famous "Light and Life" lecture:

> The recognition of the essential importance of fundamentally atomistic features in the function of living organisms is by no means sufficient, however, for a comprehensive explanation of biological phenomena, before we can reach an understanding of life on the basis of physical experience. Thus, we should doubtless kill an animal if we tried to carry the investigation of its organs so far that we could describe the role played by single atoms in vital functions. In every experiment on living organisms, there must remain an uncertainty as regards the physical conditions to which they are subjected, and the idea suggests itself that the minimal freedom we must allow the organism in this respect is just large enough to permit it, so to say, to hide its ultimate secrets from us.

It may seem strange that the numerous biological compounds in all living things, from ameoba to man, are constructed from the same twenty (or twenty-two) amino acids. The twenty-six letters of the English alphabet are enough to form all the plays of Shakespeare. The eighty-eight keys of the piano are enough for the piano concertos of Beethoven. The segregated, linear, and digital character of the genetic message is an elementary fact. Therefore, it answers the question: "What is Life" (Yockey, 1977b, 1992, 2000, 2002). There is an abyss between living organisms and inanimate matter. As Ernst Mayr (1982) put it:

> One of the properties of the genetic program is that it can supervise its own precise replication and that of other living systems such as organelles, cells and whole organisms. There is nothing exactly equivalent in inorganic nature.

The belief of mechanist-reductionists that the chemical processes in living matter do not differ in principle from those in dead matter is incorrect. There is no trace of messages determining the results of chemical reactions in inanimate matter. If genetical processes were just complicated biochemistry, the laws of mass action and thermodynamics would govern the placement of amino acids in the protein sequences.

2.2 Information as the central concept in molecular biology

Information, transcription, translation, code, redundancy, synonymous, messenger, editing, and proofreading are all appropriate terms in biology. They take their meaning from information theory (Shannon, 1948) and are not synonyms, metaphors, or analogies.

The genome is sometimes called a "blueprint" by people who have never seen a blueprint. Blueprints, no longer used, were two-dimensional, a poor metaphor indeed, for the linear and digital sequence of nucleotides in the genome. The linear structure of DNA and mRNA is often referred to as a *template*. A template is two-dimensional, it is not subject to mutations, nor can it reproduce itself. This is a poor metaphor as anyone who has used a jigsaw will be aware. One must be careful not to make a play on words.

1.2.1 Information, knowledge, and meaning

The messages conveyed by sequences of symbols sent through a communication system generally have meaning (otherwise, why are we sending them?). It often is overlooked that the meaning of a sequence of letters, if any, is arbitrary. It is determined by the natural language and is not a property of the letters or their arrangement. For example, the English word "hell" means "bright" in German, "fern" means "far," "gift" means "poison," "bald" means "soon," "Boot" means "boat," and "singe" means "sing." In French "pain" means "bread," "ballot" means a "bundle," "coin" means a "corner or a wedge," "chair" means "flesh," "cent" means "hundred," "son" means "his," "tire" means a "pull," and "ton" means "your." In French, the English word "main" means "hand," "sale" means "dirty." French-speaking visitors to English-speaking countries will be astonished at department stores having a "Sale" and especially if it is the "Main Sale." This confusion of meaning goes as far as sentences. For example, "O singe fort" has no meaning in English, although each is an English word, yet in German it means "O sing on," and in French it means "O strong monkey."

> *Meaning according to Humpty Dumpty.* When I use a word, Humpty Dumpty said in a rather scornful tone, it means just what I choose it to mean – neither more or less. "The question is," said Alice, "whether you can make word mean so many different things" "The question is," said Humpty Dumpty, "which is to be master – that's all."

Alice was too much puzzled to say anything, so after a minute Humpty Dumpty began again. "They've a temper, some of them – particularly verbs, they're the proudest – adjectives you can do anything with, but not verbs – however, I can manage the whole of them! Impenetrability! That's what I say!

Would you tell me, please said Alice "what that means"?

"Now you talk like a reasonable child," said Humpty Dumpty, looking very much pleased. "I meant by impenetrability that we have had enough of that subject, and it would be just as well if you'd mention what you mean to do next, as I suppose you don't mean to stop here all the rest of your life."

"That's a great deal to make one word mean," said Alice in a thoughtful tone.

"When I make a word do a lot of work like that," said Humpty Dumpty, "I always pay it extra."

"Oh!" said Alice. She was too much puzzled to make any other remark.

From *Through the Looking Glass*, by Lewis Carroll (1832–98), aka Reverend Charles Lutwidge Dodgson.

****

Similarly, the sequences of nucleotides or amino acids that carry a genetic message have explicit specificity. (Otherwise how does the organism live?) Now, in this book, the term *information* does not mean *knowledge*, although a message composed of a sequence of symbols may transfer knowledge to the receiver of the message.

The genetic information system operates without regard for the significance or meaning of the message, because it must be capable of handling *all* genetic messages of *all* organisms, extinct and living, as well as those not yet evolved. It does not have to be "about something."

The genetic information system is the software of life and, like the symbols in a computer, it is purely symbolic and independent of its environment. Of course, the genetic message, when expressed as a sequence of symbols, is nonmaterial but must be recorded in matter or energy. We could, in principle, send the genome of a mosquito to our little green friends on an Earth-like planet somewhere in the Milky Way Galaxy.

2

James Watson, Francis Crick, George Gamow, and the genetic code

The evidence presented supports the belief that a nucleic acid of the desoxyribose type is the fundamental unit of the transforming principle of Pneumococcus Type III.

Avery et al. *Journal Experimental Medicine* **79**, 137–159 (1943)

The phosphate-sugar backbone of our model is completely regular, but any sequence of the pairs of bases may fit into the structure. It follows that in a long molecule many different permutations are possible, and it therefore seems likely that the precise sequence of the bases is the code which carries the genetical information. If the actual order of the bases on one of the pair of chains were given, one could write down the exact order on the other one. Thus one chain is, as it were, the complement of the other, and it is this feature which suggests how the deooxyribosenucleic acid might duplicate itself.

Watson and Crick (1953b, pp. 964–5)

In a communication in Nature *of May 30, p 964, J. D. Watson and F. H. C. Crick showed that the molecule of deoxyribosenucleic acid, which can be considered as a chromosome fibre, consists of two parallel chains formed by only four different kinds of nucleotides. These are either (1) adenine, or (2) thymine, or (3) guanine, or (4) cytosine with sugar and phosphate molecules attached to them. Thus the hereditary properties of any given organism could be characterized by a long number written in a four-digital system. On the other hand, the enzymes (proteins), the composition of which must be completely determined by the deoxyribosenucleic acid molecule, are long peptide chains formed by about twenty different kinds of amino-acids, and can be considered as long 'words' based on a 20-letter alphabet. Thus the question arises about the way in which four-digital numbers can be translated into such 'words'.*

G. Gamow, *Nature*, **173**, 318 (1954a)

2.1 Watson and Crick's proposal of the role of the sequences of DNA in genetics

Those readers of this book who are computer-oriented will easily understand that the chemistry of life is controlled by digital sequences recorded in DNA, as Gamow (1954a) was the first to realize. Life is guided by information and inorganic processes are not. The publicity after fifty years still dwells on the "double helix" and biochemistry, whereas the important discovery is that the life message is digital, linear, and segregated.

The discovery by Avery and his laboratory that a nucleic acid is the fundamental carrier of the hereditary properties of life set the stage for finding the structure of DNA.

8

Watson and Crick (1953a) began their paper stating that the DNA structure published by Linus Pauling (Pauling and Corey, 1953) was wrong and proposed their own. This was incredible *chutzpah* for these two young scientists who were unknown at the time. Linus C. Pauling (1901–94) is the only person to have been awarded two unshared Nobel prizes: Chemistry (1954) and Peace (1961).

> The importance of deoxyribosenucleic acid (DNA) within living cells is undisputed. It is found in all dividing cells, largely if not entirely in the nucleus, where it is an essential constituent of the chromosomes. Many lines of evidence indicate that it is the carrier of a part of (if not all) the genetic specificity of the chromosomes and thus of the gene itself.

> The phosphate-sugar backbone of our model is completely regular, but any sequence of the pairs of bases can fit into the structure. It follows that in a long molecule many different permutations are, possible, and it therefore seems likely that the precise sequence of the bases is the code which carries the genetical information. (Watson and Crick, 1953b).

Rosalind Franklin, the dark lady of DNA. The reader may have assumed that science is carried out by selfless academics eager to give credit to all their predecessors and colleagues. That is hardly the case. Watson and Crick (1953a) cited Dr. Rosalind Franklin in a footnote of their seminal paper on the structure of DNA.

> We have also been stimulated by a knowledge of the general nature of the unpublished experimental results and ideas of Dr. M. H. F. Wilkins, Dr. R. E. Franklin and their coworkers at King's College, London.

When Watson (1968, 2001) referred to Dr. Franklin as "Rosy," "as we all called her from a distance," it was hardly a term of respect. Dr. M. H. F. Wilkins, who would share the Nobel Prize in 1962, mentioned Dr. Franklin in a footnote along with several others (Wilkins, Stokes, and Wilson, 1953). Much later, and after he had been awarded the Nobel Prize in chemistry, Watson (1968) confessed that: "Rosy, of course, did not directly give us her data. For that matter, no one at King's realized they were in our hands." But, in their first paper (Watson and Crick, 1953), they mentioned that Pauling and Corey (1953a, 1953b) had ". . . kindly made their manuscript available to us in advance of publication." They did not consider it necessary to return this courtesy or to acknowledge also that they had Dr. Franklin's data before publication.

The success of their seminal paper (Watson and Crick 1953) depended critically on their possession of the Pauling and Corey paper and on Dr. Franklin's data. Possession of both provided an opportunity to put their paper in final form and remove any mistakes. Dr. Franklin should have been a third author although she would not have been awarded the Nobel Prize in 1962. The Dark Lady of DNA died in April 1958 of ovarian cancer (Elkin, 2003; Maddox, 2002). The Nobel Prize is not awarded posthumously.

Gamow's suggestion is that life is more than complicated chemistry and that the digital information in DNA sequences is sent to the digital information in the proteome by means of a code. It was not until 2001 that Watson acknowledged the essential contribution made by George Gamow (1904–68).

2.1.1 George Gamow and his proposal of the genetic code

Gamow, promptly upon reading the paper of Watson and Crick (1953b), wrote to them proposing that the sequences of nucleotides of DNA were mapped onto the sequences of the amino acids in protein by a code (Gamow, 1954a, 1954b, 1961). Gamow's handwritten letter to Watson and Crick dated July 8, 1953, has now come to light after fifty years (Watson, 2001). Gamow wrote:

> But I am very much excited by your article in *Nature* May 30th and think that this brings biology over into the group of "exact" sciences. . . . If your point of view is correct, and I am sure it is at least in its essentials, each organism will be characterized by a long number written in quadrucal (?) system with figures 1, 2, 3, 4, standing for the four bases (or by several such numbers, one for each chromosome). It seems to me more logical to assume that different properties (single genes?) of any particular organism are not "located" in definite spots of chromosome, but are rather determined by different mathematical characters of the entire number.
> * As assumed in classical genetics

(Later at the Gatlinburg Symposium on Information Theory in Biology, I heard Gamow refer to this as "the number of the beast" [*Revelations* 13:18]: *This calls for wisdom: let him who has understanding reckon the number of the beast, for it is a human number, its number is 666.*)

Notice that Watson and Crick (1953b) used the word *code* meaning a *sequence*, whereas Gamow used the word correctly. This sloppiness is still used (Watson, 2001) and has plagued genetics and molecular biology ever since (Pennisi, 2003).

George Gamow immediately addressed the question of how the genetic specificity in the sequences selected from the four nucleotides of DNA could specify the twenty amino acids, known at that time, to be in protein (Gamow and Ycas, 1955). Samuel F. B. Morse (1791–1872), the inventor of the telegraph, faced the same problem. How can one communicate in words with only clicks on the telegraph receiver? His solution, of course, was the invention of the famous Morse Code, in which sequences of clicks and silences written as dots and dashes, separated by spaces, form an alphabet corresponding to letters and numbers.

The idea that the digital DNA sequences of four letters determined the protein sequences of twenty letters by means of a code is so unconventional in biology that had Gamow's paper been submitted by almost anyone else it would most certainly have been rejected. This was in fact the case. After he had been elected to the U.S. National Academy of Sciences, Gamow sent his paper to be published in the *Proceedings of the National Academy of Sciences*. He included his fictional friend Mr. Thompkins as coauthor. Smelling a rat, the editors returned the paper. Gamow sent the paper, without Mr. Thompkins as an author, to be published by the Royal Danish Academy, where he also had been elected a member (Gamow, 1954b). The editors of science journals had learned either to take this big, genial Ukrainian American seriously or to accept the consequences. Gamow immediately set about to find the code letters by cryptographic means (Gamow and Ycas, 1958).

When Henry Quastler (1908–63), Robert Platzman (1918–73), and I were arranging the *Gatlinburg Symposium on Information Theory in Biology*, we were delighted that Gamow accepted our invitation and also brought Martynas Ycas along. Gamow made that symposium a lively event. I was privileged to introduce him when he gave a lecture at the Oak Ridge National Laboratory. After the meeting, my wife and I drove him to the railroad station at the nearby town of Clinton. Although we corresponded frequently, that was the last I saw him. But that friendship contributed to the budding field of information theory in biology.

In his speech at the *Gatlinburg Symposium on Information Theory in Biology*, he put it this way:

> How can a sequence formed by four different units (four bases) be translated in a unique way into a sequence formed by twenty units

(twenty amino acids)? Here is a possibility which seems to us to be very likely. Suppose one plays a game of poker in which only three cards are dealt, and one pays attention only to the suit of the cards. How many different hands will one have? Well, one can have a "flush," i.e., three cards of the same suit. There are four different flushes: three hearts, three spades, etc. Then one can have as "pair," two cards of the same kind and one different. How many of those are there? One has four choices for the suit of the pair and three choices for the third card. Thus there are altogether twelve possibilities. The poorest hand will be a "bust," i.e., three different suits. There are four different "busts": no hearts, no diamonds, etc. We have altogether twenty different possibilities. This "magic number" is just the number of amino acids in the primary process of protein synthesis. We may imagine that each amino acid is determined by a triplet of bases in the RNA template. (Gamow and Ycas, 1958).

This demonstrated that a code did exist that would send information from DNA to the proteome. This code and his overlapping diamond code were soon shown to be incorrect, but the idea was fixed and the search was on.

2.1.2 *George Gamow's contributions to biology*

George Gamow was a big, tall, gregarious, and flamboyant man who was impossible to ignore. Among his first accomplishments was applying quantum mechanical tunneling through an energy barrier to explain how alpha particles, helium nuclei, could escape from the nucleus of uranium. That was an early accomplishment of quantum mechanics. But, at the time, dialectical materialism, which is incompatible with quantum mechanics, became the required theory in the Soviet Union. Gamow attempted several times to leave the Soviet Union. I remember his telling of the unsuccessful attempt he and his wife Rho made in a fold boat to cross the Black Sea. He was invited to the Solvay Conference in Belgium. The communist officials thought that an honor for the Soviet Union's scientists. Stalin, a devoted family man himself, in order to keep families together, had not allowed comrades to leave the Soviet Union with their families. Gamow's trip was approved and, of course, such an important man would need a secretary. With his wife Rho playing the role of secretary, the two arrived at the conference. Thus, his escape from *The Worker's Paradise* was like the quantum mechanical tunneling of alpha particles from uranium through an energy barrier. Communist civil servants must get up very early in the morning to outwit men like George Gamow.

He was never discouraged when one of his ideas proved wrong. His track record was such that colleagues were well aware that the next one might be correct (Segrè, 2000). He regarded science as fun and moved about at the highest levels, to whatever problem interested him. He applied his knowledge of the nascent field of nuclear physics to put the concept of the Big Bang or expanding universe theory in quantitative form.

He knew everybody in science. He founded the RNA Tie Club, and each member received a special necktie from Gamow. There were only twenty members of his RNA Tie Club, one for each of the amino acid. Those receiving special ties were: Gamow, Ala; Alexander Rich, Arg; Paul Doty, Asp; Robert Ledley, Asn; Martynas Ycas, Cys; R. Williams, Glu; Alexander Dounce, Gln; Richard Feynman, Gly; Melvin Calvin, His; N. Simons, Ile; Edward Teller, Leu; Erwin Chargaff, Lys; Nicholas Metropolis, Met; Gunther Stent, Phe; James. Watson, Pro; H. Gordon, Ser; Leslie Orgel, Thr; Max Delbrück, Tyr; Francis Crick, Trp; and Sydney Brenner, Val. This list includes five Nobel Laureates and those who were the cream of the crop in various fields at the time.

The conceptual framework, for DNA coding thus proposed by Gamow, led to the correct solution of the question of how the sequences in DNA control heredity. Without Gamow's contribution, the work may well have gone to mechanism–reductionism and perhaps dialectical materialism.

2.2 The genetic code and its relation to other codes

2.2.1 Sending the genetic message in the genome to the proteome. Just what is a code?

Therefore, for clarity, let us define what we mean by a code. Words often have different meanings that vary in the context. Lawyers speak of a code of laws. One of the earliest is the Code of Hammurabi (eighteenth century B.C.). Hammurabi, the Great King of Babylon, established these laws to bring about the rule of righteousness in the land for his loyal subjects. A systematically arranged and comprehensive collection of laws or a collection of regulations and rules of procedure or conduct may be called a code. Cryptographers say they have decoded a message. Computer programers call a code a system of symbols and rules used to represent instructions to a computer.

Following Gamow (1954a, 1961), it is essential to make a clear distinction between the *sequences* of nucleotides in the genome and the code between the mRNA alphabet and the amino acid alphabet. The *proteome*, in analogy to the genome, is the collection of amino acids that form protein. To do so, we must use a basic idea in probability theory, the *sample space*. (See discussion of probability in the Mathematical Appendix.)

> **Definition:** The set of all elementary events, *A*, in sample space
> [Ω, A. **p**] to each of which probabilities p_i have been assigned is
> called a *probability sample space of elementary events* or simply
> a *sample space*. The elements A are called *random variables*. The
> set of probabilities p_i form a probability vector **p**. The probability
> sample space is designated [Ω, *A*, **p**].

Some authors call this a *finite scheme*. According to the geometrical analogy, the elementary events in a sample space are often referred to as *points*. Examples of random variables are the number of spots on a die, the heads or tails of a coin toss, the nucleotides in DNA and mRNA, and the amino acid residues in protein.[1]

The genetic code has many of the properties of codes in general, specifically the Morse Code, the Universal Product Bar Code, ASCII [A(merican) S(tandard) C(ode for) I(nformation) I(nterchange)] used in computer equipment, and the U.S. Postal Code. I shall explain the relation of these codes to the genetic code in the following discussion. Every code, as the term is used in this book, can be regarded as a channel with an input alphabet *A* and an output alphabet *B*. Here is the formal definition of a code.

> Given a source with probability space [Ω, *A*, $\mathbf{p_A}$] and a receiver
> with probability space [Ω, *B*, $\mathbf{p_B}$], then a unique mapping of the
> letters of alphabet *A* on to the letters of alphabet *B* is called a *code*.
> (Perlwitz, Burks, and Waterman, 1988)

Here $\mathbf{p_A}$ is the probability vector of the elements of alphabet *A* and $\mathbf{p_B}$ is the probability vector of the elements of alphabet *B*.

The mathematician and philosopher René Descartes (1596–1650) showed that sample spaces, like numbers, can be multiplied (Billingsley, 1995; Feller, 1968; Khinchin, 1957; Suppes, 1972). The Cartesian product of two finite probability sample spaces Ω_0 and Ω_1, with elements *a* and *b*

---

[1] If a set of symbols is arranged in a row, they are called a *row vector*. If they are arranged in a column, they are called a *column vector*.

respectively, is the finite probability space that contains all ordered pairs (a, b). This is written $\Omega_0 \times \Omega_1$. Probability sample spaces may be raised to a power. For example, suppose Ω contains the mRNA nucleotides as random variables, elements or points, U, A, C, G. Then the sample space Ω^2 contains the ordered pairs of U, A, C, G as elements, namely, UU, UA, UC, UG, AA, AU, AC, AG, CC, CA, CG, CU, GG, GU, GA, GC, with their associated probabilities. These are, of course, the mRNA doublet codons. It follows that Ω^3 contains all the ordered triplets of U, A, C, G, with their associated probabilities, namely, the familiar mRNA triplet codons that play the role of random variables.

The primary or source alphabet used in computers and electronic communication is the binary alphabet [0, 1]. Shannon (1948) understood before anyone else that a binary source alphabet could be *extended* by forming ordered pairs, ordered triplets, ordered quadruplets, and so forth to form receiving alphabets larger than two. In computer technology, the information in the binary source alphabet is called a *bit*; these *extensions* are called a *byte*. In molecular biology these extensions are called codons. Accordingly, because of the structure of DNA and mRNA, the natural choice for the source genetic alphabet is four letters that correspond to the four nucleotides typical of DNA or mRNA.

Nature had extended the primary four-letter alphabet to the six-bit, sixty-four-member alphabet of the genetic code (Section 2.1.2). Each amino acid except Trytophan and Methionine has more than one codon. Thus, the genetic code is *redundant (not degenerate)*. The sloppy terminology designating the genetic code as *degenerate* is responsible for most of the misunderstanding of the genetic information processing system.

2.2.2 Comparison of the genetic code to other codes

The Universal Product Bar Code is attached to packaged items in stores and permits the cashier to record the price of the item. The Postal Service in the United States has established a ZIP + 4 bar code in which mailing addresses can be written. The Zip + 4 numbers are in decimal digits. The sender's computer encodes these decimal digits in a binary bar code and prints the binary bar code on the mail piece. The message in the binary Zip + 4 bar code is read by a scanner at the post office and the mail piece is sent to its destination. There the message is decoded back to decimal digits. My address is 21014-5638, and that is enough information for the postman to find the mailbox in front of my house.

Table 2.1. *The Postal ZIP + 4 code*

0	1	2	3	4	5	6	7	8	9
11000	00011	00101	00110	01001	01010	01100	10001	10010	10100
01000	10011	10101	10110	11001	11010	11100	00001	00010	00100
10000	01011	01101	01110	00001	00010	00100	11001	11010	11100
11100	00111	00001	00001	01101	01110	01000	10101	10110	10000
11010	00001	00111	00100	01011	01000	01110	10011	10000	10110
11001	00010	00100	00111	01000	01011	01101	10000	10001	10101

The alphabets of the postal and other bar codes are composed of short bars and long bars, that is, the source alphabet is binary. There are thirty-two ($2^5 = 32$) members of the fifth extension of the source binary bar code alphabet. Codes formed from these extensions are *block codes* because all code letters in the extensions have the same number of letters. The Postal Service has assigned arbitrarily the ten members of the fifth extension that have two *ones* to each of the ten digits in the decimal system. These ten members selected from the fifth extension are called *sense code letters*. Thus, the postal bar code has a five bit *byte*. The assignment of sense code letters in the fifth extension alphabet of the Postal ZIP + 4 code is shown by the first row in Table 2.1.

The other code letters are called *non-sense* because they have been given no *sense* or *meaning* assignment in the receiving alphabet. (Remember that non-sense dose not mean nonsense or foolishness.) Unfortunately, the use of the word "nonsense" persists in many current publications. I have listed those non-sense code letters of the fifth extension of the binary code with just one change from the sense code letters in each column, five rows down.

The postal ZIP + 4 code is an error-detecting code because a single error can not change one sense code letter to another sense code letter. The number of differences between code letters is called the Hamming distance (Hamming, 1986), after the originator, Dr. Richard W. Hamming. To be sophisticated, we may say that the postal ZIP + 4 code will detect but not correct any reading error that is one Hamming distance from the correct one. If, because of smudging or other malfunction, the sorting machine reads a non-sense code letter, the mail piece is rejected, to be examined by a postal employee.

Two errors are required to change one sense code letter to an incorrect sense code letter. That would result in the mail piece being sent to a wrong address. The probability of two errors is the square of the probability of one error and is therefore sufficiently small to be very rare. It is not necessary to use an error correcting code in this application.

Computer equipment uses the ASCII code, which is the seventh extension of the binary alphabet. Its *byte* is seven bits, the amount of information in one printed character in the receiver alphabet. There are $2^7 = 128$ members of the seventh extension alphabet. Members of this seventh extension of the primary alphabet are assigned arbitrarily to the letters in the receiving alphabet of the word processor. This assignment is appropriate to the language for which the word processor is designed.

Crick, Griffith, and Orgel (1957) assumed that there must be "commas" to separate a string of nucleotides forming codons to prevent them from reading out of sequence. Following the usual practice of a "magic number," they attempted to show that the maximum number of sense codons cannot be greater than twenty and gave a solution for the twenty known at the time. Unfortunately, they proved [*sic*] that the triplet AAA and other triplets must be non-sense. Table 2.2 shows that AAA codes for lysine, CCC codes for proline, UUU codes for phenylalanine, and GGG codes for glycine. Maynard Smith (1999) suggested that this was the most clever idea in the history of science that turned out to be wrong. The realization that the genetic code is a *block code* because all codons are triplets; the use of an initiator codon and codons to terminate the sequence, like the Universal Product Code and the postal Zip + 4 bar codes, makes the need for commas unnecessary.

The notion that there is a "magic number" of twenty amino acids (Weber and Miller, 1981) proved to be a red herring. A tRNA species possesses a codon complementary to UGA, one of the non-sense codons, which codes for selenocysteine, thus making the twenty-first amino acid (Burke et al., 1998; Chambers et al., 1986; Hawkes and Tappel, 1983; Lacourciere and Stadtman, 1999; Leinfelder et al., 1988; Mizutani and Hitaka, 1988, Sunde and Evenson, 1987; Zinoni et al., 1990). Paul et al. (2000) found an in-frame UAG codon in the 6.8 kb DNA in *Methanosarcina barkeri*. James et al. (2001) found that UGA, encoding monomethylamine methyltransferase isolated from *Methanosarcina barkeri*, is translated as a sense codon. Bing Hao et al. (2002) and Srinivasan et al. (2002) have found that the in-frame UAG codon is read through as L-pyrolysine, making that the twenty-second natural amino acid. There also is the question of the unnatural amino acids that I shall take up in Section 7.4.4.

Sometimes selenocysteine and L-pyrolysine are regarded as expanding the genetic code. That is a frequent misunderstanding. The genetic code is

Table 2.2. *The standard mRNA genetic code*

Amino acid	Triplet codons	Amino acid	Triplet codons	Amino acid	Triplet codons
Glycine	GGG	Phenylalanine	UUU	Leucine	UUA
	GGC		UUC		UUG*
	GGU				
	GGA				
Proline	CCG	Cysteine	UGU	Tryptophan	UGG
	CCC		UGC	Non-sense	UGA
	CCU				
	CCA				
Leucine	CUG*	Glutamine	CAA	Histidine	CAU
	CUC		CAG		CAC
	CUU				
	CUA				
Arginine	CGG	Aspartic acid	AAU	Lysine	AAA
	CGC		AAC		AAG
	CGU				
	CGA				
Threonine	ACG	Glutamic acid	GAA	Asparagine	GAU
	ACC		GAG		GAC
	ACU				
	ACA				
Valine	GUG	Isoleucine	AUU	Methionine	AUG*
	GUC		AUC		
	GUU		AUA		
	GUA				
Alanine	GCG	Non-sense	UAA	Tyrosine	UAU
	GCC		UAG (amber)		UAC
	GCU				
	GCA				
Serine	UCG			Arginine	AGA
	UCC				AGG
	UCU			Serine	AGU
	UCA				AGC

Purines: Adenine, A; Guanine, G
Pyrimidines: Uracil, U; Cytosine, C
* Initiator codons

fixed at sixty-four codons. These additional amino acids expand the genetic *alphabet*.

2.2.3 The genetic code

The genetic code is a mapping of the mRNA code letters in the genome on to the code letters of the proteome. It is not merely a table of correlations. The source alphabet of the genetic code is the quaternary alphabet of four nucleotides of DNA and mRNA. Thus each nucleotide has a two-bit byte. The first extension of that quaternary alphabet has sixteen letters and a four-bit byte. However, that is not sufficient to code the canonical twenty amino acids that are transcribed in protein and to provide for the starting and stopping function corresponding to the long bars at the ends of the Postal Zip + 4 bar code. Accordingly, Nature has gone to the second extension sixty-four-letter alphabet. Thus, the genetic code has a six-bit byte, called a codon or in computer technology a code word.

The mRNA genetic code, shown in Table 2.2, shares a number of properties with the Postal ZIP + 4 code, the ASCII computer codes, and the Universal Product Code. The genetic code is *distinct* and *uniquely decodable*, because the single Methionine codon AUG, and sometimes the Leucine codons UUG and CUG, serve as a starting signal for the protein sequence and performs the same function as the long frame bars at the beginning of the postal message in the ZIP + 4 code and the Universal Product Code. The codons UGA, UAA and UAG function usually as non-sense and stop the translation of the protein from the mRNA and initiate the release of the protein sequence from the mRNA (Maeshiro and Kimura, 1998). They perform the same function as the long frame bar at the end of the postal bar code message (Bertram, 2001).

3

The Central Dogma of molecular biology

Ah, but my Computations, People say,
Have squared the year to human compass, eh?
If so by striking from the calendar
Unknown tomorrow and dead Yesterday.
The Rubaiyat of Omar Khayyam (Fitzgerald, Second Edition)

3.1 Francis Crick and the Central Dogma

Francis Crick (1958) published *The Central Dogma*, stating his view of how DNA, mRNA and protein interact. The Central Dogma states that information can be transferred from DNA to DNA, DNA to mRNA and mRNA to protein. Three transfers that the Central Dogma states *never* occur are protein to protein, protein to DNA, protein to mRNA.

> On the other hand, the discovery of just one type of present day cell which could carry out any of the three unknown transfers would shake the whole intellectual basis of molecular biology, and it is for this reason that the central dogma is as important as when first proposed. (Crick, 1970)

Crick need not have worried. He emphasized, correctly, that there is no flow of matter, but, rather, ". . . sequence information from one polymer molecule to another." I wrote to Professor Crick (private correspondence, 2002) congratulating him on the Central Dogma. He replied that he believed that the Central Dogma is only an hypothesis. I have shown long ago that Professor Crick hath wrought better than he knew (Yockey 1974, 1978, 1992, 2002).

3.1.1 The Shannon entropy criterion for codes that transfer messages in one alphabet to another

The genetic code has a Central Dogma because it is *redundant*. As a result, except for Trytophan and Methionine, it is *undecidable* which source code letter was actually sent from mRNA. The Central Dogma, stated correctly, is a mathematical property of *any* computing or information processing system that uses a redundant code. It is not a fundamental property of the chemistry of nucleic acids and amino acids (Yockey, 1974, 1978, 1981, 1995a, 1995b, 1992, 2000, 2002a, 2002b). Two alphabets are isomorphic, *if and only if*, they have the same Shannon entropy (Billingsley, 1965; Kolmogorov, 1958; Ornstein, 1970, 1974; Shields, 1974). The Shannon entropy of the DNA alphabet and the mRNA alphabet is $\log_2 64$ (Section 4.1). The Shannon entropy of the proteome alphabet is $\log_2 20$; thus, like all codes between sequences that are not isomorphic, the genetic code has a Central Dogma. No code exists that allows information to be transferred from protein sequences to mRNA. Therefore, it is *impossible* for the origin of life to be "proteins first" (Yockey, 1992, 2000, 2002a, 2000c).

Furthermore, the Central Dogma reflects the well-known biological fact that acquired characteristics cannot be inherited (Battail, 2001; Yockey, 1974. 1978, 1992, 1995, 2000a, 1995, 2002a). Thus, evolution can only be Darwinian.

The restrictions of the Central Dogma on the origin of life are mathematical (Battail, 2001; Yockey, 1974, 1978, 1992, 2000, 2002a). Scientists cannot get around them by clever chemistry. Likewise, Nature's proscription against the building of perpetual motion machines is also mathematical. The Second Law of Thermodynamics places a severe limit on the ability of a clever engineer to build machines that derive work from heat. Regardless of the choice of materials or design it is *impossible* to build a perpetual motion machine. These restrictions apply however socially, politically, and environmentally desirable it may be to make perpetual motion machines. (See the distinction between *impossible* and *unknowable* in Chapter 11.)

3.1.2 Reverse transcription and my reappraisal of Crick's Central Dogma

Baltimore (1970) and Temin and Mizutani (1970) reported, independently, that RNA tumor viruses contain an enzyme that uses viral RNA to transcribe the sequence of DNA. The viral-specific enzyme *reverse transcriptase* catalyses the transcription of the HIV genome retroviral infection into a complementary mRNA sequence. An unsigned article in *Nature* (1970) interpreted the work of Baltimore (1970) and that of Temin and Mizutani (1970) as requiring a critical reappraisal of the Central Dogma. Crick (1970) replied that: "It (the Central Dogma) was intended to apply only to present-day organisms, and not to events in the remote past, such as the origin of life or the origin of the code." That is too modest an evaluation of the Central Dogma.

It is obvious that if the source and receiver alphabets have the same Shannon entropy, information may be passed without loss, in either direction. Thus, because mRNA and DNA both have a four-letter alphabet, a one-to-one correspondence can be established so that genetic messages may be passed from mRNA to DNA, or from DNA to mRNA. Thus, the genetics obeys the mathematics. The so-called *reverse transcription* that astonished many people is in accordance with the discussion of the theory of codes in general in Chapter 2, Chapter 7, and in Section 3.1.1 (Yockey, 1974, 1978, 1992, 2000, 2002). David Baltimore and Howard Temin (Baltimore, 1970; Temin and Mizutani, 1970) were awarded the Nobel Prize for Physiology or Medicine in 1975 for the discovery of *reverse transcription*.

Reversible computation has been well studied by computer engineers, Landauer (2000), Bennett (1973, 1988), Bennett and Landauer (1985), and Zurek (1984, 1989). There would have been very much less *Sturm und Drang* in the olive groves of academe if that knowledge had been applied to understanding the genetic code.

3.1.3 Misunderstandings of the Central Dogma

The Central Dogma was widely misunderstood at the beginning (Commoner, 1964, 1968; unsigned article, 1970) and is still so today (Henikoff, 2002; and an unsigned article in *Nature Genetics*, 2002). Commoner (1968) showed that he did not understand the subject because he called the "Watson–Crick Theory" a *chemical explanation* of inheritance. He is not alone in that mistake today.

Harpers Magazine is known for its sophisticated political commentary, and that is where Commoner (2002), a longtime leftist radical social and environmental activist, published his attack on genetic engineering. His choice of *Harpers Magazine* shows that his purpose is purely political. He points out the horrible achievements of genetic engineering:

> Pigs now carry a gene for bovine growth hormone and show significant improvement in weight gain, feed efficiency, and reduced fat. Most soybean plants grown in the United States have been genetically engineered to survive the application of powerful herbicides. Corn plants (maize) now contain a bacterial gene that produces an insecticide protein rendering them poisonous to earworms.

These scare tactics have prevented the acceptance of desperately needed food in parts of Africa affected by a drought-induced famine.

He presents the Human Genome Project as some sort of capitalist plot involving the genetic information system . . . *a serpent in the biotech garden* (*Genesis* 3:1–7). Using language appropriate to leftist political jargon, he wrote:

> The wonders of genetic science are all founded on the discovery of the DNA double helix by Francis Crick and James Watson in 1953 and they proceed from the premise that this molecular structure is the exclusive agent of inheritance in all living things: in the kingdom of molecular genetics, the DNA gene is absolute monarch. Known to molecular biologists as the "central dogma" the premise assumes that an organism's genome – its total complement of DNA genes – should fully account for its characteristic assemblage of inherited traits. The premise, unhappily is false.

Commoner confuses the *Central Dogma* with the *Sequence Hypothesis* (Section 2.1.1).

Commoner believes that a fatal fault in Human Genome Project is that there are too few human genes to account for the complexity of our inherited traits or for the vast inherited differences between plants, say, and people. On the contrary, there are many genes common to all living things, to perform common functions, such as the formation of the twenty amino acids found in all life. Commoner commits a particular blooper when he writes:

> Because of their commitment to an obsolete theory, molecular biologists operate under the assumption that DNA is the secret of life, whereas the careful observation of the hierarcharchy of living processes strongly suggests that it is the other way around: DNA did not create life: life created DNA.
> [*Now he really steps on his argument:*]
> When life was first formed on the Earth, proteins must have appeared before DNA because, unlike DNA, proteins have the catalytic ability to generate the chemical energy needed to assemble small ambient molecules into large ones such as DNA.

This quotation shows again that Commoner does not understand the genetic information system and the Central Dogma. It is mathematically *impossible*, not just *unlikely*, for information to be transferred from the protein alphabet to the mRNA alphabet. That is because no codes exist to transfer information from the twenty-letter protein alphabet to the sixty-four-letter alphabet of mRNA.

Response to Commoner's article may be found on the Web site <http://www.criticalgenetics.org>. These responses make it quite clear that the Central Dogma is widely misunderstood.

3.2 Prions

Crick (1970) mentioned the disease scrapie specifically as a possible exception to the restriction of the transfer of information, protein-protein. The chemical agent of scrapie, called a *prion*, is a proteinaceous infective agent devoid of nucleic acid (Griffith, 1967; Kimberlin, 1982; Liebman, 2002; Maddelein et al., 2002; Prusiner, 1982, 1998; Wills, 1986, 1989; Peretz et al., 2001). Prions seem to be composed exclusively of a modified isoform of Prp designated PrP^{Sc}.

Protein-protein recognition plays a central role in most biological processes (Kortemme and Baker, 2002). All biological processes are interdependent so that many proteins have several cellular roles. There is no mathematical restriction of the transfer of information from protein to protein. If a protein-protein genetic code were to exist, that is allowed by the Shannon entropy theorems and would be no violation of the Central Dogma.

3.3 Energy dissipation due to computation in the genetic logic operation and its relation to the Central Dogma

Energy use by the cell is priced in the sense that, should such losses become too large, it diminishes the energy available for use by the cell for other needs. Therefore, the energy use of the cell places a limit on the errors in the genetic message that can be corrected. Nevertheless, some genetic errors can be tolerated and, consequently, perfect accuracy is neither necessary nor desirable because the ability to evolve depends on some flexibility in the genetic message carried by the DNA. This situation is much the same as the complexity problems in computer data processing systems (Section 2.4.3), in which the data also is subject to error and the complexity needed to correct error is priced in the terms of computer memory and computer time.

Although there is no analog to energy in communication theory, nevertheless, the actual equipment used by the communication engineer dissipates energy and the biological communication system does likewise (Bennett and

Landauer, 1985; Landauer, 1986, 2000). In each case, the question arises: What is the minimum energy dissipation required by the basic physics of the operation? Both systems are subject to thermal noise because of the indeterminancy in the laws of quantum mechanics. Kullback (1959) defined the mutual information $I(2|1)$ in statistical inference as the average information per observation for discrimination in favor of observation H_2 against H_1 as:

$$I(2|1) = \int f_2(X) \log_e \left(\frac{f(X)}{f_1(X)} \right) d\,\lambda(x), \tag{3.1}$$

where $\lambda(x)$ is a probability measure. From this equation, Kullback (1959) rigorously derived the correct equation for the channel capacity of a continuous signal:

$$I(\Delta\bar{\omega}, \Delta t) = (\Delta\bar{\omega} \times \Delta t) \log_e \left(1 + \frac{S}{P} \right), \tag{3.2}$$

where S is the signal, power (energy per unit time), P is the thermal noise power, $\Delta\bar{\omega}$ is the frequency band width of a continuous signal, and Δt is the duration. Equation 3.2 also had been obtained by Shannon (1948) but without the same rigor. P is given by:

$$P = kT\Delta\bar{\omega}, \tag{3.3}$$

where k is the Boltzmann constant and T is the absolute temperature. In our case, the ratio $\frac{S}{P}$ is small compared to one. Accordingly, the logarithmic expression, $\log_e (1 + x)$, in Equation 3.2 may be replaced approximately by x. If one makes that substitution and some rearrangement, one finds the following equation for the minimum energy per bit:

$$\frac{S\Delta t}{I(\Delta\bar{\omega}, \Delta t)} = kT \log_e 2 = 0.693\,kT. \tag{3.4}$$

The ratio of the signal energy $S\,\Delta t$ to the number of bits in the signal, if the alphabet is binary, is $0.693\ kT$. Of interest to molecular biology, the discrimination is between four observations and the logarithm of four must appear. The energy per bit is $1.386\ kT$ or about 0.035 electron volts. This dissipation of energy occurs at each step in the genetic logic operation, including proofreading. Because the ratio of signal power of thermal noise power is very small, the expression on the right of Equation 3.4 is the minimum energy per bit. This result also had been obtained by Shannon (1948) and later by von Neumann (1966).

The question of the minimum dissipation of energy in computation and communication has received considerable attention in the theory of computers, because the engineering problem of heat removal limits the compactness and therefore the speed of computing machinery. The basic process in the calculations leading to Equation 3.4 is that noise discards information and consequently the communication process is irreversible. Nevertheless, DNA transcription and translation may be the closest approach to a Brownian computer dissipating 100–20 kT per step (Bennett, 1973).

In the case of the genetic logic system, it is clear that information is discarded by the redundant genetic codons between mRNA and protein as well as by genetic noise. Therefore, the process dissipates energy, generates Maxwell–Boltzmann–Gibbs entropy, and is irreversible. There is no *élan vital* in biology, so the genetic logic system must obey the same fundamental laws as other logic operations. There is considerable discussion in the literature that leads to the result that an ideal computer can operate without a limit on energy dissipation if it operates slowly enough and if the computations leading to the output are not erased. Neither of these conditions applies to the genetic logic system. Once the tRNA has accomplished its coding operation, except in the case of Met and Trp, the information with regard to the exact codon is erased from the memory of the genetic logic system. The considerations given in this section of the question of energy dissipation in genetic logic operations are sufficient for the purposes of this book, but the reader may wish to consult significant papers in the field, namely, Landauer (2000), Bennett (1973, 1988), Bennett and Landauer (1985), and Zurek (1984, 1989).

4

The measure of the information content in the genetic message

The road not taken

Two roads diverged in a yellow wood.
And sorry I could not travel both
And be one traveler, long I stood
And looked down one as far as I could
To where it bent in the undergrowth;
Then took the other as just as fair,
And having perhaps the better claim,
Because it was grassy and wanted wear;
Though as far that the passing there
Had worn them really about the same,
And both that morning equally lay
In leaves no step had trodden back.
Oh, I kept the first for anoter day!
Yet knowing how way leads to way,
I doubted if I should ever come back,
I shall be telling this with a sigh
Somewhere ages and ages hence;
Two roads diverged in the wood, and I-
I took the one less traveled by,
And that has made all the difference.
                                    Robert Frost (1874–1963)

4.1 The measure of the information in the genetic message: The road not taken

It is almost universally believed that the number of sequences in polypeptide chains of length N, composed of the twenty common amino acids that form protein, can be calculated by the following expression:

$$(20)^N. \tag{4.1}$$

Expression 4.1 gives the total number of sequences we must be concerned with *if and only if all events are equally probable*. However, many events in general, and amino acids in particular, do not have the same probability. Unfortunately, many distinguished authors have neglected that fact and led their readers and students astray. (See MA1.1, *The Origins and Interpretation of Probability*, in the Mathematical Appendix.)

27

But let us take the road less traveled; it will make all the difference and lead to the correct way to calculate the number of sequences in a family of nucleic acid and polypeptide chains. Shannon (1948) addressed this problem as follows: Let us consider a long sequence of N symbols selected from an alphabet of A symbols. In the present case, the symbols will be the alphabet of either codons or amino acids. Just as in the toss of dice, there is no intersymbol influence in the formation of these sequences. Let $p(i)$ be the probability of the ith symbol. The sequence will contain $Np(i)$ of the ith symbol. Let P be the probability of the sequence. Then, because the probabilities of independent events are multiplied (Shannon, 1948):

$$P = \prod_i^N p(i)^{p(i)N}. \tag{4.2}$$

Taking the logarithm of both sides changes multiplication to addition:

$$\log_2 P = N \sum_i p(i) \log_2 p(i) \tag{4.3}$$

$$\log_2 P = N \sum_i p(i) \log_2 p(i) = -NH, \tag{4.4}$$

where

$$H = - \sum_i p(i) \log_2 p(i), \tag{4.5}$$

H is called the Shannon entropy of the sequence of events.[*]

In communication, genetics, and molecular biology, we are interested in long sequences. Accordingly, the probability of a long sequence of N independent symbols or events taken from a finite alphabet is:

$$P = 2^{-NH}. \tag{4.6}$$

The number of sequences of length N is:

$$2^{NH}. \tag{4.7}$$

Notice that the expression for H was not introduced *ad hoc*; rather, it comes out of the woodwork, so to speak.

---

[*] Some authors are confused by the minus sign in Equation 4.5 and that leads them to believe in negative entropy (see Section 4.4). The probabilities of all events being considered must sum to 1. Probabilities lie between zero and one. Logarithms in that range are negative or zero, so $\log_2 P$ is always zero or negative and so are the terms $\log_2 p(i)$. We always take $0 \log 0$ to be zero. Therefore, Shannon entropy is always positive or zero.

Logarithms to base 2 can be calculated by the use of a pocket calculator: $\log_2 y = \log_{10} y / \log_{10} 2$.

Let us compare Expression 4.1 and Expression 4.7 by calculating the number of sequences in one hundred throws of a a pair of dice, where the probabilities of all events are known exactly and are not all equal. For a given throw, the probability of 2 and 12 is 1/36, whereas the probability of 7 is 6/36 because there are six ways to roll a 7 and only one way to roll 2 or 12. So we see that the number of sequences calculated by Equation 7 is only 2.69×10^{-6} of that calculated from expression (1), namely, 11^N.

We have calculated the number of sequences of length N in two apparently correct ways and the question arises: What happened to the sequences left out by the second method? This is explained by the Shannon–McMillan– Breiman Theorem (Breiman, 1957; McMillan, 1953; Shannon, 1948):

> For sequences of length N being sufficiently long, all sequences being chosen from an alphabet of A symbols, the ensemble of sequences can be divided into two groups such that:
> 1. The probability P of any sequences in the first group is equal to 2^{-NH}
> 2. The sum of the probabilities of all sequences in the second group is less than ε, a very small number.

The Shannon–McMillan–Breiman Theorem is a surprising result. It tells us that the number of sequences in the first, or high, probability group is 2^{NH} and they are all nearly equally probable. We can ignore all those in the second or low probability group because, if N is large, their **total probability** is very small. The number of sequences in the high probability group is almost always many orders of magnitude smaller than that given by Expression 1, which contains an enormous number of "junk" sequences. In a fast-forward to Section 6.4, I find that the information content of 1-iso-cytochrome c, a small protein of 113 amino acids is 233.19 bits. The number of 1-iso-cytochrome c sequences is $6.42392495176 \times 10^{111}$. Calculating this number by Expression 1, we find $20^{113} = 1.03845927171 \times 10^{147}$. The 1-iso-cytochrome c sequences are only a very tiny fraction, $6.1968577266 \times 10^{-36}$ of the total possible sequences. Thus, one sees that Expression 1 is extremely misleading.

One must further remember that the word *entropy* is the name of a mathematical function, nothing more. One must not ascribe meaning to the function that is not in the mathematics. For example, the word *information* in this book is never used to connote knowledge or any other dictionary meaning of the word *information* not specifically stated here.

The road we have taken, the one less traveled, has led us to the Shannon–McMillan–Breiman Theorem. It is, almost without exception, unknown to authors in molecular biology, and without it they have been led to many false conclusions. As in the sequences of throws of a pair of dice, all DNA, mRNA, and protein sequences are in the high probability group and are a very tiny fraction of the total possible number of such sequences.

I seem to be the only one to have applied the Shannon–McMillan–Breiman Theorem in molecular biology [Yockey, 1974, 1977, 1981, 2000, 2002a, 2002c]. This will lead to my comment in Chapter 10 on Eigen's proposal of a "master sequence" in a "quasi-species" each one having a "value parameter" or "superiority parameter" (Eigen 1971, 1992, 2002).

(The proof of the Shannon–McMillan–Breiman Theorem is in the Mathematical Appendix.)

4.2 Shannon entropy as a measure of information, uncertainty, randomness, choice, and ignorance

First, as Socrates taught us, we must establish a measure of the *information* in a message. Like all messages, the life message has a measurable information content. Readers who are computer literate are familiar with H being measured in *bits*, when one takes the logarithm to base 2.

To use Shannon entropy as a measure of uncertainty and choice, we must first say about what we are uncertain. Suppose our uncertainty is the outcome of the football or baseball season, and we wish to choose the winner of the season. Without apology, we may establish a probability for each team, because these probability measures are our personal degree of belief. We only are required that the sum of all probabilities be equal to one. As the season progresses, some teams will lose games, while others will be winners. Then, one may adjust the appraisal of each team's probability of winning according to the record. At the end of the last game, the winner has prevailed, one is no longer uncertain of the winner, and Shannon entropy goes to zero (see Shannon, 1948, para. 6: *Choice, Uncertainty and Entropy*).

4.3 Entropy in probability theory and entropy in statistical mechanics

Many authors have been misled by the resemblance of Equation 5 to that for entropy in statistical mechanics (Brillouin, 1953, 1962, 1990; Chaisson, 2001; McDermott, 2002). The Second Law of Thermodynamics is appropriate to the design of heat engines but has nothing to do with evolution. The probability sample space in classical statistical mechanics, called *phase space* by theoretical physicists, is six-dimensional, and the probabilities are defined by the position and momentum vectors of the particles in the ensemble. The function for entropy in both classical statistical mechanics and the von Neumann entropy of quantum statistical mechanics has the dimensions of the Boltzmann constant k and has to do with energy and momentum, not information (Petz, 2001; von Neumann, 1932). Entropy in information theory and probability theory has no mechanical dimensions. There are no counterparts in communication theory to temperature, energy, pressure, work, or volume. There is, furthermore, no counterpart to the First Law of Thermodynamics, namely, the conservation of the energy of a system.

To illustrate this point further, one may consider the probability space of a dice game that consists of the numbers two through twelve as random variables and calculate the corresponding entropy. Clearly, the Shannon entropy of a dice game has nothing whatever to do with statistical mechanics or thermodynamics. It may have something to do with information theory, as a series of tosses of two dice selects a sequence of symbols from the alphabet two through twelve. Such a sequence forms a message in which some gamblers find meaning or knowledge by which they make their bets. By contrast, information theory is concerned with messages expressed in sequences of letters selected from a finite alphabet. The letters of the alphabet construct the sample space under consideration as random variables and the p_i are defined accordingly.

4.4 The question of negentropy: Can entropy be negative?

The great mathematician Norbert Wiener (1894–1964) regarded negative entropy as a measure of information (Wiener 1948). Simpson (1964) had the following remark:

> A fully living system must be capable of energy conversion in such a way as to accumulate *negentropy*, that is, it must produce a less probable, less random organization of matter and must cause the increase of available energy in the local system rather than the decrease demanded in closed systems by the Second Law of Thermodynamics.

Brillouin (1953, 1962) used the concept also to address certain problems in physics and in 1990 to address the relation of thermodynamics and life.

Schrödinger (1987, 1992) used negative entropy to explain the appearance of what he thought was *order* during evolution. Eigen (1992) thought that information received is negative entropy. It is most unfortunate that these distinguished scientists have misled their readers. Perhaps they believe that the minus sign in Equation 5 means that Shannon entropy is negative entropy. Because all probabilities must range from zero to one the logarithm is negative and that means that H is zero or positive. That is elementary mathematics. This is a serious mathematical objection to the *ad hoc* notion of negentropy in addition to the fact that a means of measurement has not been proposed (Khinchin, 1957; Yockey, 1977, 1992).

These distinguished authors are confusing Shannon entropy of probability theory with Maxwell–Boltzmann–Gibbs entropy of statistical mechanics. Contrary to Schrödinger (1987, 1992), Wiener (1948), Eigen (1992), and a number of authors, whose name is Legion for they are many, Shannon entropy is not *negentropy*. Life does not feed on *negentropy* (Pauling, 1987) as a cat laps up cream. The notion of *negentropy* has crept into textbooks and the technical and popular literature. It must be exorcised to avoid more damage.

5

Communication of information from the genome to the proteome

The fundamental problem of communication is that of reproducing at one point either exactly or approximately a message selected at another point. Frequently the messages have meaning; that is they refer to or are correlated according to some system with certain physical or conceptual entities. These semantic aspects of communication are irrelevant to the engineering [biological] problem. The significant aspect is that the actual message is one selected from a set of possible messages. The system must be designed to operate for each possible selection, not just the one which will actually be chosen since this is unknown at the time of design.

Shannon (1948)

5.1 Genetics and the standard communication system

5.1.1 The components of the genetic communication system

Let us now consider the model shown in Figure 5.1 of a general communication system commonly used by communication engineers. The object of such systems is to accept messages from the source and to transmit them through a channel to the destination as free from errors as the specifications given to the design engineer require. The source generates an ensemble of messages written in the finite source alphabet, *A*. The message is encoded from the source alphabet to the channel alphabet for transmission through the channel. At all stages of the communication the message is acted on by a second chance or stochastic process (see Markov process in The Mathematical Appendix) that interchanges some letters in a random and nonreproducible fashion. The result of this process is called *noise*. It occurs in all blocks, but it is lumped for clarity in Figure 5.1. The ensemble of messages, modified by noise, is received and decoded to the alphabet *B* at the destination.

5.1.2 The DNA-mRNA-proteome communication system in genetics

Figure 5.2 describes the DNA-mRNA-proteome communication system to show its isomorphism with the standard communication system of the communication engineer. The genome, or the ensemble of genetic messages, is generated by a stationary Markov process and recorded in the DNA

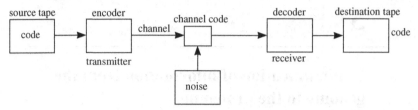

Figure 5.1. The transmission of information from source to destination as conceived in electrical engineering. Noise occurs in all stages but is shown according to accepted practice in electrical engineering.

sequence, which is isomorphic with the tape in a tape-recording machine (Turing, 1936).

The decoding of the genetic message from the DNA alphabet to the mRNA alphabet is called *transcription* in molecular biology. mRNA plays the role of the channel, which communicates the genetic message to the ribosomes, which serve as the decoder. The genetic message is decoded by the ribosomes from the sixty-four-letter mRNA alphabet to the twenty-letter alphabet of the proteome. This decoding process is called *translation* in molecular biology. Figures 5.3a and 5.3b show actual electron micrographs of the ribosomes as they move along the mRNA sequence decoding from the mRNA alphabet to the protein alphabet and thereby producing protein (Kiseleva, 1989). They act like the reading head on a tape machine (Turing, 1936). The protein molecule, which is the destination, is also a tape. Thus, the one-dimensional genetic message is recorded in a sequence of amino acids, which folds up to become a three-dimensional active protein molecule. One is reminded of the linear signals that fold up to show a two-dimensional picture on the television screen.

The direction of flow of information is governed by the Shannon entropy of the alphabets where encoding and decoding take place. Thus, if a redundant code is used in any system described in either Figure 5.1 or Figure 5.2, that system has a Central Dogma and information flows only from the source to the destination (see Chapter 3). In the retroviral case, where the two alphabets have the same Shannon entropy, information may flow in either direction in the DNA-mRNA encoding if the process is catalyzed by a reverse transcriptase (see Section 3.1.3).

This process has evolved to meet the requirements of all organisms that ever lived, those that are alive today and all organisms yet to evolve.

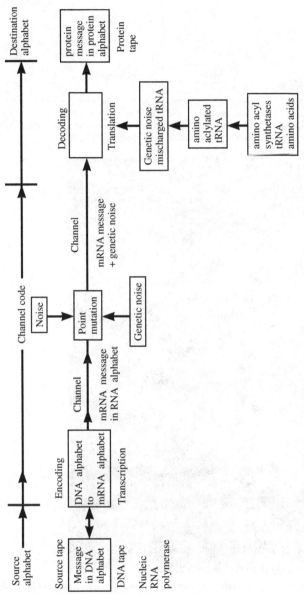

Figure 5.2. The transmission of genetic messages from the DNA tape to the protein tape as conceived in molecular biology. Genetic noise occurs in all stages but is lumped in the figure to fix the idea.

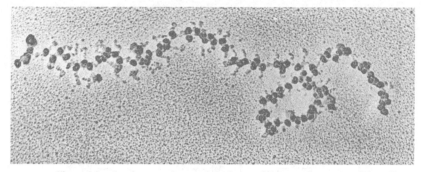

Figure 5.3a. An electron micrograph is shown of the spread contents of the salivary glands of *Chironomus thummi*. Single and double arrows indicate the 5′ and 3′ ends of mRNA, respectively. The individual polyribosomes and endoplasmic membrane-bound polyribosome complexes are isolated from the secretory cells.

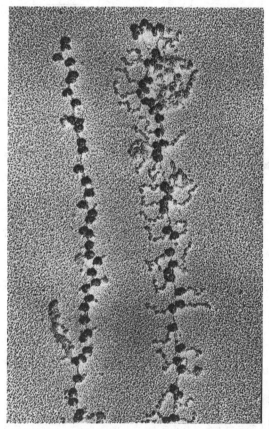

Figure 5.3b. Shows an electron micrograph of the ribosomes acting like the reading head on a tape machine (Turing, 1936), as they move along the mRNA sequence, decoding from the mRNA alphabet to the protein alphabet and thereby producing protein. From Kiseleva (1989) with permission.

5.1.3 The definition of genetic noise

The source that generates the message is regarded mathematically as a stochastic Markov process characterized by a probability space $[\Omega, A, \mathbf{p}]$ where A, the alphabet of the source, like the spots on dice, is a random variable, the value of which is assumed *after* each toss. Ideally, the sequences generated by the source in alphabet A will be those recorded in alphabet B at the receiver. In the real case there is always a second random Markov process that acts on the original message so that a letter in the original message may cross over to another letter at the receiver. This has been called *missense*, and that like the terms *nonsense* and *degeneracy*, is an unjustifiably pejorative survivor of the days when the genetic code was not understood. In Chapter 6, I discuss the fact that many sites in a protein sequence may be occupied by a number of functionally equivalent amino acids. Mutations to a functionally equivalent amino acid do not affect the activity of the protein. Thus, such mutations are not *missense* or errors.

Let us borrow a term from communication engineering and call this effect *genetic noise* (Yockey, 1956, 1958, 1974, 1977b, 1992, 2000, 2002a). This effect is lumped in Figure 5.2, but it operates at each of the components of the communication system. In molecular biology, this process is caused by point mutations. If the interchange (crossover) of letters is made with equal probability, the stochastic process is called *white noise*. I shall call this phenomenon *white genetic noise* if the probability of a random change between any two nucleotides is equal. This is analogous to the practice in communication engineering.

White noise is an idealization that will seldom be found in the laboratory, so, in general, the effect of point mutations will be called simply *genetic noise*. Nevertheless, white noise has important applications in coding theory. In practice, the transition matrix of the Markov process that operates on the message from the source is not known (see Markov Process and the Random Walk and Section MA1-3.1 in the Mathematical Appendix). Some forms of genetic noise are not white noise; for example, burst errors such as may be caused by a flaw in a computer tape or by an alpha particle passing through a segment of DNA. Although burst error correcting codes have been constructed in communication theory, they are highly specialized. Insertions and deletions cause a frame shift that results in a *catastrophic error propagation*. This phenomenon also is known to communication engineers and is to be avoided in the construction of communication codes. Almost all considerations of noise in coding theory and information theory are directed at white noise.

5.1.4 The role of the majority logic redundancy of gene duplication in error correction

Sending the message several times to overcome errors is called *majority logic redundancy*. The receiver is faced with a collection of messages that may have errors scattered about and few if any may be identical. This method of error correction is appropriate if there is some means to check accuracy. For example, suppose the message is the combination of a safe. We may try each of the several messages received and if one is correct the safe will open. There may be duplicate combinations that will open the safe. It can be shown that the probability of error may be made as small as we wish by this method, but then the transmission rate is reduced to zero (Ash, 1965).

Duplicated genes play an essential role in evolution as Susumu Ohno (1929–2000) proposed (Ohno, 1970). Eukaryotic genomes have 8–20 percent duplicates. The rate of gene duplication is estimated to be between 0.2 percent and 2 percent per gene per million years (Moore and Purugganan, 2003). Gene redundancy has been observed in yeast, plant, and human genomes (Adams et al. 2003; Bowers et al. 2003; Langkjaer et al. 2003). The duplicated genes provide redundance protection against mutation to a nonspecific message but also to the evolution of a new gene (Graure and Li, 2000; Haldane 1932). Gu et al. (2003) estimate that in the yeast *Saccharomyces cerevisiare* genome at least a quarter of these gene deletions that have no phenotype are compensated by duplicate genes. Current results in genetic sequencing show that duplicated genes are abundant in most genomes (Lynch 2002; Lynch and Conery, 2000).

There is considerable redundance in 1-iso-cytochrome c and presumably in other proteins, as I shall show in Table 6.3. That table shows that there are $6.42392495176 \times 10^{111}$ 1-iso-cytochrome sequences in the high probability set. Presumably there also are an enormous number of sequences of other proteins. Majority logic serves in genetics to receive at least one of this enormous number of functional protein sequences in the proteome, whereas there is only one correct sequence received in telecommunications. Contrary to Eigen's "error catastrophe" (Grande-Perez et al. 2002; Eigen 2002), I shall show in Chapter 12 that because of this redundance, the genetic message can indeed survive for 3.85 billion years from the origin of life.

5.2 Error-detecting and error correcting codes

5.2.1 Parity checks

Parity checks have long been used in evaluating the accuracy of telecommunication messages, especially when the message is in financial transactions. When the message is sent in binary alphabet [1, 0] the simplest method is to count the number of 1s and to append a final binary digit so that the entire message has an even number of 1s. An odd number of 1s in the message received indicates that there is at least one error. Any odd number of errors can be detected by parity checks. Of course, this is not much help, because we don't know where the error is. This method is useful only for short messages. These methods of error detection and correction are long out of date for commercial messages and for messages to and from spacecraft.

5.2.2 Error detecting and error correcting properties of block codes

Because the genetic code is a block code, like the Postal ZIP + 4 code (Section 2.2.2), let us consider how block codes, constructed from the nth extension of the source code letters, are used to construct error-detecting and -correcting codes. In the nth extension of the alphabet A, there are a total of q^n sequences of code letters. Where q is the number of alphabetical symbols. Of these $q^k (k \leq n)$ are a subset that is used by the source to send messages to the receiver. These sequences are called *sense code letters*. For example, suppose the source alphabet is binary and the fifth extension is used so there are five digits in each code letter and thirty-two possible code letters. Let the alphabet, B, at the receiver be a quaternary code, like the genetic code U, A, C, G. Of the thirty-two possible code letters, a subset of at least four *sense code letters* in alphabet A is needed to be mapped onto the receiver alphabet B. Suppose the source and the receiver agree that the source sense code letters are those listed at the head of each column of Table 5.1. A receiver of primitive construction would decode only the source sense code letters to the alphabet B. For this primitive receiver, the twenty-eight code letters that are not sense code letters are *non-sense* code letters.

The codons UAA, UGA, and UAG are usually *non-sense* codons. Referring to them as *nonsense* codons is not correct. Because they are not ordinarily assigned code words in the receiver alphabet they act as chain terminators and initiate the release of the protein sequence from the mRNA. As a matter of fact, as I shall show in Chapter 7, they do have assignments

Table 5.1. *Error correcting block code: decoding table q = k = 2 and n = 5*

Code words in receiver alphabet B	U	A	G	C
Sense code words in source alphabet A	11000	00110	10011	01101
Code words received at Hamming	11001	00111	10010	01100
distance 1	11010	00100	10001	01111
	11100	00010	10111	01001
	10000	01110	11011	00101
	01000	10110	00011	11101
Code words received at Hamming	11110	00000	01011	10101
distance 2	01010	10100	11111	00001

From *Error Correcting Codes* (Second Edition) by Peterson & Weldon (1972) with permission from the Mit Press.

in mitochondria to code selenocysteine and L-pyrolysine, making those the twenty-first and twenty-second natural amino acids (Section 2.2.2). Source code letters that are assigned to the same destination code letter are called *synonymous*. Notice that there are no *ambiguous* codes. That is, each of the source code words is always assigned to only one destination code letter, although, as in the genetic code, there may be more than one source code words assigned to a code word in the receiving alphabet.

5.2.3 The effect of the hamming distance in error-correcting and -detecting codes

The number of positions in which synonymous source code letters differ is called the *Hamming distance* (Hamming, 1950). The sense code letters listed at the head of each column in Table 5.1 were selected such that they have the largest possible Hamming distance between them and each code letter in the other columns. Five of the synonymous code letters in each column are separated from the sense code letters by only one Hamming unit. The two code letters at the bottom of each column are separated by two Hamming units. In each column, all the source sense code letters with only one error, because of white noise, and two that have two errors, are received correctly. It is possible to decode all patterns of t or fewer errors, if and only if, the minimum Hamming distance is at least $2t + 1$. All other cases result in the sense code letter being decoded incorrectly. Of course, the source could be perverse and send any of the synonymous code letters in each column. This would compromise the error-detecting and -correcting property for which the code was designed. Referring, for example, to the code in Table 5.1, clearly, any four of the thirty-two source code letters could have been chosen as sense code letters. It is undecidable at the destination which of the set of correct code letters was actually sent. If the source does stick to the sense code letters, the decoder may obtain a measure of the noise on the line and still decode correctly if the noise does not become too great.

In this example, the decoder also could be designed to choose a second extension of the quaternary alphabet to include all pairs of the U, A, C, G, and thus sixteen of the source code letters would be sense code letters. However, then $k = 4$, and only two other source code letters would be received correctly. The cost of this would be a greater vulnerability to white noise. Furthermore, in the limit, each of the thirty-two code letters could be designated as a sense code letter. In that case, determining which of the set of correct code letters was actually sent is decidable, but there is no protection at all from white noise. If thirty-two code letters were needed, the designer of the code would have to go to a sixth and presumably a seventh extension to construct an error-correcting and error-detecting code.

If all the source code letters are mapped on the destination alphabet so that the source entropy and the destination entropy are equal, the code is said to be *complete* or *saturated*. Thus, the DNA-mRNA code is complete or saturated, whereas the mRNA-protein code is not saturated.

5.2.4 *The error correction effect of redundancy in the genetic code*

One of the ways to make a code more secure from white noise is to assign more than one alternate source code letter according to the frequency of code letters in the destination alphabet. Such a code uses the redundancy in a nearly optimum manner to reduce the effect of genetic noise (Figureau and Labouygues, 1981). The best-protected codons are Leu (CUA, CUG), and Arg (CGA, CGG). The least-protected is UGG for Trp, which has two transitions to a termination codon.

It has been pointed out by many authors for several years that the genetic code is arranged to minimize the effect of genetic noise. First, these authors recognized that, having assigned one codon to each of the twenty amino acids, some assignment of the remaining forty-four codons, other than non-sense or stop, was necessary to reduce the vulnerability to mutation, that is, to genetic noise. As this process evolved, the genetic code became nearly *complete* or *saturated* in the form we find it today. This is clearly illustrated by analogy with the error correcting binary code in Table 5.1. Thus, *redundance* in the genetic code plays a role in contributing to error protection.

The designation of the genetic code as *degenerate* (Edelman and Gally, 2001; Tononi, Sporns, and Edelman, 1999) rather than *redundant* shows that these authors do not understand the role of redundance in the genetic code. The use of this confusing terminology in the early papers still haunts molecular biology to its detriment.

The evidence is very plausible, both from the biochemistry of protein and the mathematical requirements of the code for error correction, that the present code assignment resulted from selection pressure very early in the history of life (Chapter 7). The modern genetic code is optimal (Section 7.4.2) for the twenty amino acids for which codons have been assigned and it could not be improved without an extension to four or more letters in each codon (Cullmann and Labouygues, 1985). Apparently, this is not possible considering the structure of DNA. Thus, the fact that more than one codon is assigned to eighteen of the more common amino acids in protein is seen as very natural, and indeed necessary, to achieve a moderate error-correcting capability in the genetic code.

5.3 Shannon's Channel Capacity Theorem

Although the genetic system is remarkably accurate, genetic noise (Yockey, 1956, 1958, 1974, 1977b, 2000, 2002) does cause some codons to be translated or decoded incorrectly. One might be tempted to give up the possibility of correct transmission if it were not for an unexpected contribution from communication theory. Shannon's Channel Capacity Theorem for a noisy channel states that a code exists between alphabet A and alphabet B such that the communication system can transmit information as close to the channel capacity as one desires with an arbitrarily small error rate. This is done at the cost of incorporating redundance in an error correcting code. By contrast, it is not possible to transmit at a rate greater than the channel capacity.

The theorem is not constructive and is no help in finding such codes. However, this and other theorems establish from first principles *conditional entropy* (Section 2.1.4) as a *measure* of the effect of noise and, in general, *mutual entropy*, as a *measure* of the relationship between any two sequences one wishes to compare (Section 5.2.2). *Remember that in this book we deal only with those concepts that can be measured.*

There are several ways that Nature has incorporated redundance for error correction in the genetic code and in the protein message. I discuss these in Chapters 6, 7, and 11. We may therefore proceed to apply Shannon's Channel Capacity Theorem to problems in molecular biology, secure in the knowledge that we are not just inventing an *ad hoc* procedure.

5.3.1 The measurable properties of the source and the channel

A *discrete memoryless source* is defined as one in which there are no restrictions or intersymbol influences between letters of the alphabet such as *qu* in English. A channel that allows input symbols to be transmitted in any sequence is called an *unconstrained channel*. A source that transmits messages written in natural languages is not a *memoryless source*, as natural languages do have intersymbol influence. Thus, the DNA-mRNA-protein system is *discrete, memoryless*, and *unconstrained*. The particular message recorded in the DNA is independent of the genetic information apparatus.

In communication theory, the messages that have *meaning*, or in molecular biology *specificity*, are imbedded in the ensemble of random sequences that have the same statistical structure, that is, the same Shannon entropy. We know the statistical structure of the ensemble but not that of the individual sequences. For that reason, the output of any information source, and, in particular, DNA in molecular biology, is regarded as a random process that is completely characterized by the probability spaces $[\Omega, A, \mathbf{p_A}], [\Omega, B, \mathbf{p_B}]$. The alphabets A and B are random variables. Shannon considered only sources with the character of stationary Markov processes. This was generalized to include other sources by McMillan (1953). For our purposes, at least at first, we may regard the particular message transmitted as being one member of a stochastic ensemble generated by a stationary Markov process. The code letters of the source, the channel, and the destination are the states of a Markov process.

5.3.2 Conditional entropy is the proper measure of genetic noise

Shannon (1948, para. 12: *Equivocation and Channel Capacity*) gave an elementary and very lucid explanation of the proper measure of noise and how that measure should be used to calculate the amount of information that can be transmitted through a noisy channel. This explanation is repeated here with permission and a change of notation to conform with that in this book.

> Suppose there are two possible symbols 0 and 1, and we are transmitting at at rate of 1000 symbols per second with probabilities $p_0 = p_1 = 1/2$. Thus our source is producing information at a rate of 1000 bits per second. During the transmission the noise introduces errors so that, on the average 1 in 100 is received incorrectly (a 0 as 1, or 1 as 0). What is the rate of transmission of information? Certainly less than 1000 bits per second since about 1% of the received symbols are incorrect. Our first impulse might be to say that the rate is 990 bits per second, merely subtracting the number of errors. This is not satisfactory since it fails to take into account the recipient's lack of knowledge of where the errors occur. We may carry it to an extreme case and suppose the noise is so great that the received symbols are entirely independent of the transmitted symbols. The probability of receiving a 1 is 1/2 whatever was transmitted and similarly for 0. Then about half of the received symbols are correct due to chance alone, and we would be giving the system the credit for transmitting 500 bits per second while actually no information is being transmitted at all. Equally "good" transmission would be obtained by dispensing with the channel entirely and flipping a coin at the receiving point.
>
> Evidently the proper correction to apply to the amount of information transmitted is the amount of this information that is missing in the received signal, or alternatively the uncertainty when we have received a signal of what was actually sent. From our previous discussion of entropy as a measure of uncertainty it seems reasonable to use the conditional entropy of the message, knowing the received signal, as a measure of this missing information. This is indeed the proper definition, as we shall see later. Following this idea the rate of actual transmission, R, would be obtained by subtracting from the rate of production (i.e., the rate of events x at the source) the average rate of conditional entropy (of events, y, at the receiver)
>
> $R = H(x) - H(x \mid y).$

The conditional entropy $H(x \mid y)$ will, for convenience, be called the equivocation. It measures the average ambiguity of the received signal.

In the example considered above, if a 0 is received, the *a posteriori* probability that a 0 was transmitted is 0.99, and that a 1 was transmitted is 0.01. These figures are reversed if a 1 is received. Hence:

$$H(x \mid y) = -[0.99 \log_2 0.99 + 0.01 \log_2 0.01]$$
$$= 0.081 \text{ bits per symbol}$$

or 81 bits per second. We may say that the system is transmitting at a rate $1000 - 81 = 919$ bits per second. In the extreme case where a 0 is equally likely to be received as a 0 or a 1 and similarly for 1, the *a posteriori* probabilities are 1/2 and 1/2 and

$$H(x \mid y) = -[1/2 \log_2 1/2 + 1/2 \log_2 1/2]$$
$$= 1 \text{ bit per symbol}$$

or 1000 bits per second. The rate of transmission is then 0 as it should be.

This anecdotal explanation shows that the probabilities of the errors *cannot* be a measure of the amount of information that is lost by noise. It shows that it is plausible that conditional entropy is the proper measure. It is also apparent that R in the quotation above is a measure of the shared or mutual information of two sequences. The measure of this mutual information is called the *mutual entropy*. This conjecture will be reinforced by the discussion of the mathematical properties in Section 5.2.2. For example, I show in Theorem 5.1 that the mutual entropy is symmetric between two sequences and in Theorem 5.2 that it is zero, if and only if, the two sequences are independent. (See the discussion of mutual entropy in the Mathematical Appendix.) The application to specific problems in molecular biology is given in Section 5.4 and in Chapter 6.

We are always allowed to make definitions, provided we also prove theorems that show that the new concept is not merely a convenient empiricism. Definitions are useful if, and only if, such theorems exist, otherwise they are empty. Shannon was too good a mathematician to leave the matter as given in the earlier quotation. He proved that conditional entropy is the proper measure of the effect of noise. In addition, he proved a number of powerful theorems about mutual entropy that I shall discuss in Section 5.4.

5.4 The properties of mutual entropy

Let us now consider the use of mutual entropy as a measure of the information transmitted from source to receiver. (The proof will be found in the Mathematical Appendix.) The input or source of the channel is characterized by a probability space $[\Omega, A, \mathbf{p_A}]$ with an input alphabet A of elements x that compose probability vector $\mathbf{p_A}$. The output of the channel is characterized by a probability space $[\Omega, B, \mathbf{p_B}]$ with an output alphabet B of elements y. There is a conditional probability matrix, $\mathbf{P}$, with matrix elements, $p(j \mid i)$, which gives the probability that if letter y_j appears at the output, then the letter x_i was sent (see Section MA1.3.1). The probabilities p_i and p_j are related by the following equation:

$$p_j = \sum_i^n p_i p(j \mid i). \tag{5.1}$$

Definition The *mutual entropy* describing the relation between the input of the channel and the output is defined as follows:

$$I(A; B) = H(x) - H(x \mid y), \tag{5.2}$$

where $H(x \mid y)$ is the conditional entropy of x_i given that y_j has been received. The conditional entropy is written in terms of the components of probability vector $\mathbf{p}$ and the elements of probability matrix $\mathbf{P}$ as follows:[*]

$$H(x \mid y) = -\sum_{ij} p_j p(i \mid j) \log_2 p(i \mid j). \tag{5.3}$$

Theorem 5.1 The value of the mutual entropy is symmetric between the source and the receiver $I(A; B) = I(B; A)$.

Theorem 5.2 The mutual information $I(A; B)$ is zero, if and only if, the sequences in alphabet A and those in alphabet B are independent.

---

[*] In some publications Equation 5.2 is written:

$$I(i) = H_{before} - H_{after}$$

That is incorrect. Papers in which that appears should be disregarded.
Adami, Ofria, & Collier (2000).
Adami & Cerf (2000).
Schneider, Thomas D. (2000).
Schneider, Stormo, Gold, & Ehrenfeucht (1986).

(The proofs of Theorem 5.1 and 5.2 can be found in the Mathematical Appendix.)

Mutual information is a measure of the information in one set of sequences about another (Chaitin, 1975; Shannon, 1948). One can think of mutual entropy as a measure of the information that the output of a channel, with an alphabet B, a random variable, gives about the input from the source with another random variable, namely, alphabet A. Consequently, the mutual entropy is a measure of the *similarity* or *resemblance* of the sequences in alphabet A to those in alphabet B and is essentially the mathematical expression of what we mean by *similarity* or *resemblance*. Thus, words must take their meaning from the mathematics, not the other way around. If the sequences in alphabet A and those in alphabet B are identical then $H(x \mid y)$ vanishes. Therefore, all values of the mutual entropy lie between $H(x)$ and zero.

Because $I(A; B)$ has a finite maximum value, it is reasonable, in the case of a communication channel, to call that maximum the *channel capacity*. According to Shannon's Channel Capacity Theorem, there is no code that can be used to transmit information at a rate greater than the channel capacity. This is intuitively reasonable, as sequences in alphabet B cannot be more similar to sequences in alphabet A than identity.

We are, in fact, not limited in this interpretation of mutual entropy and Shannon's Channel Capacity Theorem. We may remove the communication system entirely and consider the mutual entropy between any two sequences in alphabets A and B. Mutual entropy is a measure of their *similarity*. Thus, the irrelevant and uncorrelated details of each set of sequences cancel out, revealing their *similarity*. These theorems establish mutual entropy as the *only* measure of *similarity* between sequences and avoids the assumption of knowledge we do not have.

5.5 The mutual entropy of homologous protein families

5.5.1 The distinction between "homologous" and "similar"

The study of natural proteins shows that they are grouped in homologous families, for example, myoglobin, the α and β hemoglobin chains and iso-1-cytochrome c. These sequences perform similar functions in the organisms in which they are found. The term "homologous" means having a common evolutionary origin as evidenced by common function and structure.

Thus, the term "homologous," like the word "unique," cannot be qualified. Sequences are either homologous or they are not. The term "similar or resemblance" with which "homologous" is often confused, means "being identical except for a number of sites in the chain." Human and gorilla iso-1-cytochrome c are both very similar and homologous, but human and yeast iso-1-cytochrome cc are homologous but less similar. Thus, because sequences may be more or less similar the "similarity" of the sequences can be quantified.

5.5.2 Mutual entropy must replace "percent identity" as a measure of the similarity of amino acid or nucleotide sequences: Reinventing the wheel

There is a need in molecular biology to have a measure of the similarity between sequences or the information in one set of sequences about another. The notion prevails in the olive groves of academe that the appropriate measure of the similarity, or sometimes the degree of homology between sequences is an *ad hoc* score, namely, the *percent identity* in the amino acid alignment (Doolittle, 1981, 1987a, 1987b, 1988). This is a case of reinventing the wheel. Mutual entropy has been available since 1948. *Percent identity* is not a measure of similarity for the same reasons that error frequency is not a measure of the effect of noise on the transmission of information (Section 5.2.1).

In addition, the percent identity score does not take into account the fact that the letters of the alphabet of the sequences are, almost always, not equally probable, nor, in the application to molecular biology, does it take into account the degree of functional equivalence between the amino acid replacements. Thus, two sequences may have a small percent identity score yet be closely related. I have pointed out previously (Yockey, 1974) that *which* amino acids may occupy a site as well as *how many* is important. It will be clear from the following discussion that the mutual entropy, as calculated from Equation 5.2, takes the functional equivalence of amino acid replacements and their probability into account, without *ad hoc* assumptions and, accordingly, is the only measure of similarity that is universally quantifiable (Section 5.2.2).

Similarity derives its meaning and mathematical definition from mutual entropy, just as other words find their meaning in mathematical definitions and the theorems that justify the definitions (Chapter 2). Mutual entropy is the measure of the *similarity* of any two sequences *or set of sequences* of whatever kind, whether or not they are associated with communication systems and however they may occur (Section 5.2.2). In contrast, the *ad hoc* percent identity score cannot be applied to more than two sequences. This is a more general statement than the one in the first application of mutual entropy in establishing Shannon's Channel Capacity Theorem.

Mutual entropy is symmetrical between a sequence from a source and a sequence at a receiver (Theorem 5.1). If the sequences are identical, the mutual entropy has its maximum value. If the sequences are independent, the mutual entropy is zero (Theorem 5.2). If the sequences are those of nucleotides in DNA, RNA or protein, then mutual entropy is a measure of biological similarity. Because homology cannot be qualified (i.e., two or more sequences are either homologous or they are not), the sloppy expression "degree of homology" must be replaced by mutual entropy as a measure and definition of similarity. Two or more sequences may have a large mutual entropy as a result of *convergent* evolution. The criterion for best alignment is that which exhibits the most similarity. Consequently, the best alignment is that for which the mutual entropy is at its maximum. No *ad hoc* corrections need be made to allow for chance coincidences, as the equations that define mutual entropy account for chance coincidences.

5.4.1 *Genetic noise expressed as mutual entropy of the genome*

It will prove to be more convenient to deal with the message at the source and therefore with the probability vector elements, p_i, and the matrix elements, $p(j \mid i)$ (Yockey, 1974). (See Section MA1.3 in the Mathematical Appendix.) From Bayes' Theorem on conditional probabilities we have Section 1.2.6 (Feller, 1968; Hamming, 1986; Lindley, 1965):

$$p(i \mid j) = \frac{p_i \, p(j \mid i)}{p_j}, \tag{5.4}$$

where the $p(j \mid i)$ matrix elements are the forward conditional probabilities and the $p(i \mid j)$ matrix elements are the backward conditional probabilities. Substituting this in the expression for $H(x \mid y)$ in Equation 5.2 we have:

$$I(A; B) = H(x) - H(y \mid x) - \sum_{i,j} p_i \, p(j \mid i) \, [\log_2(p_j / p_i)], \tag{5.5}$$

Table 5.2. *Genetic code transition probability matrix elements* $p(j/i)^a$

Amino acid y_j Codon x_i	Leu	Ser	Arg	Ala	Val	Pro	Thr	Gly	Ile	Term	Tyr	His	Gln	Asn	Lys	Asp	Glu	Cys	Phe	Trp	Met
UUA	$(1-7\alpha)$	α			α				α	2α									2α		
UUG	$(1-7\alpha)$	α			α					α									2α	α	α
CUU	$(1-6\alpha)$		α		α	α			α			α							α		
CUC	$(1-6\alpha)$		α		α	α			α			α							α		
CUA	$(1-5\alpha)$		α		α	α			α				α								
CUG	$(1-5\alpha)$		α		α	α							α								α
UCU		$(1-6\alpha)$		α		α	α				α							α	α		
UCC		$(1-6\alpha)$		α		α	α				α							α	α		
UCA	α	$(1-6\alpha)$		α		α	α			2α											
UCG	α	$(1-6\alpha)$		α		α	α			α										α	
AGU		$(1-8\alpha)$	3α				α	α	α					α				α			
AGC		$(1-8\alpha)$	3α				α	α	α					α				α			
CGU	α	α	$(1-6\alpha)$			α		α				α						α			
CGC	α	α	$(1-6\alpha)$			α		α				α						α			
CGA	α		$(1-5\alpha)$			α		α		α			α								
CGG	α		$(1-5\alpha)$			α		α					α							α	
AGG		2α	$(1-7\alpha)$				α	α							α					α	α
AGA		2α	$(1-7\alpha)$				α	α	α	α					α						
GCU		α		$(1-6\alpha)$	α	α	α	α								α					
GCC		α		$(1-6\alpha)$	α	α	α	α								α					
GCA		α		$(1-6\alpha)$	α	α	α	α									α				
GCG		α		$(1-6\alpha)$	α	α	α	α									α				
GUU	α			α	$(1-6\alpha)$			α	α							α			α		
GUC	α			α	$(1-6\alpha)$			α	α							α			α		
GUA	2α			α	$(1-6\alpha)$			α	α								α				
GUG	2α			α	$(1-6\alpha)$			α									α				α
CCU	α	α	α	α		$(1-6\alpha)$	α					α									
CCC	α	α	α	α		$(1-6\alpha)$	α					α									
CCA	α	α	α	α		$(1-6\alpha)$	α						α								
CCG	α	α	α	α		$(1-6\alpha)$	α						α								
ACU		2α		α		α	$(1-6\alpha)$		α					α							

a Reformatted from Yockey, H.P. An application of information theory to the central dogma and the sequence hypothesis, *Journal of Theoretical Biology* (1974): 46: 369–406. Published with permission.

ACC	2α	α	α															α
ACA	α	α	α	$(1-6\alpha)$														
ACG	α	α	α	$(1-6\alpha)$														
GGU	α	α	α	$(1-6\alpha)$													α	
GGA	α	2α	α	$(1-6\alpha)$														
GGC	α	α	α	$(1-6\alpha)$												α	α	α
GGG	2α	α	α	$(1-6\alpha)$												α	α	α
AUU	α	α	$(1-7\alpha)$															
AUC	α	α	$(1-7\alpha)$															
AUA	2α	α	$(1-7\alpha)$													α	α	
UAA	α	α	$(1-7\alpha)$		2α													
UAG	α	α	$(1-8\alpha)$		2α											α	α	
UGA	α	2α	$(1-8\alpha)$	α												α	α	α
UAU	α	α			2α	2α	$(1-8\alpha)$								2α	α		
UAC	α	α			2α	2α	$(1-8\alpha)$								α	α		
CAU	α	α			$(1-8\alpha)$	$(1-8\alpha)$	α											
CAC	α	α			$(1-8\alpha)$	$(1-8\alpha)$	α											
CAA	α	α			2α	2α	α	α							α			
CAG	α	α			2α	2α	α	α							α			
AAU	α	α	α		α	α		$(1-8\alpha)$	$(1-8\alpha)$									
AAC	α	α	α		α	α		$(1-8\alpha)$	$(1-8\alpha)$									
AAA	α	α		α	α	α	α	2α	2α								α	
AAG	α		α	α	α	α	α	2α	2α									
GAU	α	α	α				$(1-8\alpha)$	$(1-8\alpha)$										
GAC	α	α	α				$(1-8\alpha)$	$(1-8\alpha)$										
GAA	α	α	α				2α	2α	α									
GAG	α	α	α				2α	2α	α	α								
UGU	2α	α		α	α						$(1-8\alpha)$	α						
UGC	2α	α		α	α						$(1-8\alpha)$	α						
UUU	3α	α	α								$(1-8\alpha)$	α						
UUC	3α	α	α								$(1-8\alpha)$	α						
UGG	α	2α	α		2α						α	2α	$(1-9\alpha)$	$(1-9\alpha)$				
AUG	2α	α	α	3α			α						$(1-9\alpha)$					

where

$$H(y \mid x) = - \sum_{i,j} p_i p(j \mid i) \log_2 p(j \mid i). \qquad (5.6)$$

$H(y \mid x)$ vanishes if there is no genetic noise because then the matrix elements $p(j \mid i)$ are either 0 or 1 (remember that $0 \log 0 = 0$). The third term in Equation 5.5 is the information that cannot be transmitted to the receiver if the source alphabet is larger than the alphabet at the receiver, that is, if the Shannon entropy of the source is greater than that of the receiver so that the source and the receiver alphabets are not isomorphic. It is true in general and it is a manifestation and quantitative measure of the effect of the Central Dogma that was discussed in Chapter 3.

Let us first consider the genetic noise caused by mischarged tRNA species. We may divide the mischarged tRNA species into two groups; (1), those in which the codon of the mischarged amino acid differs by only one nucleotide from the appropriate codon in the mRNA and (2) those in which more than one nucleotide differs from the appropriate codon. Misreading one nucleotide is much more likely than misreading two. If we assume this is true in general, we can set up the matrix elements of the transition probability matrix **P**. I have done this in Table 5.2 (Yockey, 1974, 1992), in which α is the probability of a base interchange of any one nucleotide, all interchanges being equally probable. By lumping all these probabilities in a single parameter, we are calculating the effect of *white genetic noise*. Substitute the matrix elements from Table 5.2 in Equations 5.5 and 5.6 and, replacing the logarithm by its expansion including only terms of the second degree, we have (Yockey, 1974, 1992):

$$I(A; B) = H(\mathbf{p}) - 1.7915 - 9.815\alpha$$
$$+ 34.2108\alpha^2 + 6.8303\alpha \, \log_2 \alpha. \qquad (5.7)$$

In the absence of noise, the terms in α vanish and the number 1.7915 is the difference in Shannon entropy of the source and the receiver and is, therefore, the amount of information that cannot be transmitted from an mRNA sequence to a protein sequence because of the redundance in the genetic code (remember that $0 \log 0 = 0$). There are actually twelve transversion and transition probabilities among the four nucleotides considering both directions of change. If these transversion and transition probabilities were known, an equation similar to Equation 5.7 can be derived with these additional parameters. However, Equation 5.7 is sufficient to describe the general effect of genetic noise on the genome.

Whether or not the amino acids replaced by the action of genetic noise are errors in the protein sequence must be determined at each site in a particular protein. To get the actual decrease in mutual information as a function of α, one must substitute 0 for α if the replacement amino acid is functionally acceptable at *each site* in the protein sequence. As α increases, the value of the mutual information falls nearer and nearer to that of the information content of the protein being considered. Some protein sequences are functionally active and some are not. Consequently, the population of functional proteins falls gradually below the level needed to preserve the viability of the cell (Yockey, 1958b, 1992, 2000, 2002) (see Chapter 10).

5.4.2 Mutual entropy as a measure of information content or complexity of protein families

In Section 6.2, I discuss the functionally equivalent replacements that may be made at each site in iso-1-cytochrome c. If one selects an active iso-1-cytochrome c from the ensemble of all iso-1-cytochrome c sequences, one is uncertain which of the several functionally equivalent amino acids occupies any given variable site. Clearly, the measure of this uncertainty is the conditional entropy $H(y \mid x)$ given in equation 5.4 and 5.5. We may therefore subtract the conditional entropy, $H(y \mid x)$ from the source entropy. This will give us a measure of the information content at that site. If the site is invariant, there is no uncertainty and the conditional entropy vanishes: The alphabet of the source is larger than that of the receiver and, consequently, the entropy of the source is larger than that of the receiver. In order to take this difference into account, it is more instructive to use Equations 5.4 and 5.5.

We must now find the conditional probability matrix **P** of the Markov chain that describes the evolution communication channel (Cullmann and Labouygues, 1987). The conditional probability matrix must be obtained in its equilibrium state. The conditional probability matrix **P**, with matrix elements $p(j \mid i)$, completely describes the communication channel and the probability of the appearance of the codons of an amino acid residue, j, following the occurrence of a codon i. As shown in Table 6.3, as many as nineteen amino acids may appear at certain sites in iso-1-cytochrome c.

The method of calculating the value of the matrix elements follows from the discussion of the Perron–Frobenius Theorem in *The Mathematical Appendix*. We now divide the codons into two groups. The codons for invariant amino acids and codons for those amino acids that are not functionally equivalent at the site in question. They will obey the genetic code; therefore, the matrix elements will be either zero or one. There is no uncertainty and those terms in Equation 5.4 will vanish. We now may allow the functionally equivalent amino acids to mutate among themselves. The matrix elements of **P** are the transition probabilities for the codons of the functionally acceptable amino acids at a given site. **P** must be doubly stochastic and regular. (We recall that a regular matrix is one in which, at some power, all matrix elements are > 0; see the Mathematical Appendix). Let $\mathbf{p}_0$ be the prior probability vector of the functionally equivalent amino acids at a given site in the protein sequence. Let t be the number of steps in which a mutation is fixed in a population. Let us remind ourselves that $\mathbf{P}^t$ is a λ-matrix (see the Mathematical Appendix), where the elements are polynomials of degree t. We can, if we wish, stop at any step t to calculate the matrix elements and vector components for substitution in Equations 5.13 and 5.14. We can, as a matter of fact, follow the progress of the mutual entropy, step by step, to its equilibrium value.

P is a square doubly stochastic matrix, because the nucleotides interchange among themselves. We may therefore raise **P** to the power t. The probability vector after t steps will be $\mathbf{p}_t$

$$\mathbf{P}^t \mathbf{p}_0 = \mathbf{p}_t. \tag{5.7}$$

As t grows beyond bounds, the matrix elements approach those of the limiting transition matrix **T**. In the limit they become equal to each other and all knowledge of the original probability vector $\mathbf{p}_0$ is lost. Therefore, as time goes on, eventually the $p(j \mid i)$ can all be set equal to $1/s$, where s is the total number of codons of all the functionally equivalent amino acids at a given site. We shall divide the amino acids into classes $C_6, C_4, C_3, C_2,$ and C_1, the subscripts indicating the number of codons for each class. Here I shall assume that the probability of each amino acid is proportional to the number of codons. The number of codons for the class of the functionally equivalent residue j is r_j. Recalling Equation 5.1, for all functionally equivalent amino acids we have

$$p_j = \sum_j (1/n)(1/s) r_j = \sum_j r_j (1/ns) \tag{5.8}$$

and therefore, for the functionally equivalent amino acids *only* where $\sum r_j = r$:

$$p_j = r/ns. \tag{5.9}$$

The sum of all codons pertaining to the class of the accepted amino acids is r. That is, if Ser and Tyr are the functionally equivalent amino acids, then $r = 6 + 2 = 8$. We may now substitute in Equations 5.4 and 5.5

$$I = H(\mathbf{p}) + \sum_{i,j} \left(\frac{1}{n}\right)\left(\frac{1}{s}\right) \log_2\left(\frac{1}{s}\right)$$

$$- \sum_{i,j} \left\{ \left(\frac{1}{n}\right)\left(\frac{1}{s}\right) \log_2\left[\left(\frac{r}{ns}\right)n\right] + \left(\frac{1}{n}\right) \log_2 r_j \right\}. \quad (5.10)$$

Combine the first terms of the second and third expressions in Equation 5.2:

$$I = H(\mathbf{p}) + \sum_{i,j} 1/n[(1/s) \log_2(1/s) - (1/s) \log_2(1/s)$$

$$- (1/s) \log_2 r] - \sum_{i,j}(1/n)\log_2 r_j. \quad (5.11)$$

The first two terms in the second term in the brackets of Equation 5.11 cancel. There are $s \times r$ terms in the third expression in the bracket, so that after performing the summation that term becomes $-(r/n)\log_2 r$. The last term is summed over the amino acids that are *not* included in the class of the functionally accepted mutations. Upon summation, that term produces terms such that the coefficient of $\log_2 r_j$ is the number of amino acids, a_j, in class C_j *not* included in the accepted ones, multiplied by the number of codons for that amino acid. Then Equation 5.11 reduces to the following very simple equation:

$$I = \log_2 n - (r/n) \log_2 r - a_6(6/n)\log_2 6 - a_4(8/n)$$

$$- a_3(3/n) \log_2 3 - a_2(2/n). \quad (5.12)$$

Equation 5.12 takes into account the information not needed when there is more than one functionally equivalent amino acid residue, the probability of those amino acids, and the information that cannot be transmitted to protein because of the redundancy of the genetic code. The more amino acids that are functionally equivalent at a given site in an homologous protein family, the less information is needed to specify at least one such residue.

In deriving Equation 5.12, I have assumed the amino acid probabilities are proportional to the number of codon assignments in the genetic code. The probabilities p_j are often not proportional to the number of codons in the genetic code and in those cases this must be taken into account. Jukes, Holmquist, and Moise (1975) suggested the following proportions:

$Ala_{5.3}Arg_{2.7}Asn_{3.2}Asp_{3.3}Cys_{1.4}Gln_{3.2}Glu_{3.3}Gly_{4.9}His_{1.3}Ile_{3.0}Leu_{3.9}Lys_{3.9}$
$Met_{1.1}Trp_{0.8}Tyr_{2.0}Val_{4.1}$. One may establish as many as twenty classes and assign a number of fictitious codons, according to these subscripted numbers r_j to each class (Yockey, 1977b). The r_j need not be whole numbers. Equation 5.22 takes the following form

$$I = \log_2 n - (r/n)\log_2 r - \sum_j \delta_j a_j (r_j/n)\ \log_2 r_j \qquad (5.13)$$

$$r = \sum_j (1 - \delta_j)r_j, \qquad (5.14)$$

where $\delta_j = 0$, if the jth amino acid is included in the set of functionally equivalent amino acids, and $\delta_j = 1$ if not.

The Perron–Frobenius Theorem is not well known, but without it I would not have been able to find Equation 5.12. Thus, in Equation 5.12, we have the correct means to calculate the *information content* or the *complexity* of a *family* of protein sequences (Yockey, 1992). I shall do this in Chapter 6. We recall that Equation 4.7 provides the means to calculate the number of sequences that have information content H. That calculation may well be the most important in this book. It will lead us to the conclusions in Chapters 10, 11, and 12 on the questions of how much is knowable about evolution and the origin of life.

6

The information content or complexity
of protein families

We have only begun to appreciate the tremendous amount of biological information implicit in the biochemistry of living organisms.

M. O. Dayhoff and R. V. Eck (1978)

6.1 The information content or complexity of an homologous protein family

6.1.1 Functionally equivalent amino acids

The specificity of proteins is determined, not only by the amino acid sequence, but also by the active pocket of amino acids that contains metal ions such as iron, zinc, copper and manganese (Thompson and Orvig, 2003). Thyroxine, which contains iodine, is the major hormone secreted by the thyroid gland. Thyroid gland deficiency disease is very common, especially in women.

Some substitutions of amino acids at certain sites may have a destabilizing effect on the protein-folding pathways. Thus, the selectivity of amino acids is determined by the primary role played in the protein folding process as well as by the requirements of the activity of the completed and folded molecule (Hoang et al., 2002). Proteins that misfold can form extracellular or intracellular aggregates, resulting in disastrous cellular dysfunction. Human protein-folding disorders include Alzheimer's and Parkinson's diseases (Selkoe, 2003).

6.1.2 The sequence hypothesis of Watson and Crick and Shannon's information theory

We now are able to address the application of Shannon's information theory to the sequence hypothesis of Watson and Crick. Usually in the olive groves of academe, suggestions from other departments are not received gladly. Nevertheless, Gamow's proposal that the sequence hypothesis could bring biology over into the group of the exact sciences could not be ignored.

Communication systems are concerned with sending messages from here to there, from past to present, or from the present to the future. Let us consider evolution as a communication system from past to present and from present to future. As an example, take cytochrome c, a small globular protein heme-containing an iron ion, formed early in the evolution of life. It is an essential protein and performs a key step in the production of the energy of the cell. So, in dealing with iso-1-cytochrome c we are examining the essence of the metabolism of all living cells. The c-type cytochromes have a long history. Almasy and Dickerson (1978) trace the cytochrome c super family to the earliest fermenting bacteria. For the time being, let us take 3.85 billion years as a working date for the appearance of life (Mojzsis and Harrison, 2002; Mojzsis, Kishnamurthy, and Arrhenius, 1999). Kunisawa et al. (1987) suggest that the cytochrome c superfamily can be traced back 3.2×10^9 years. Baba et al. (1981) date the origin of eukaryotic cytochrome c at 1.4×10^9 years ago. Wu et al. (1986) estimate a figure of 1.2×10^9 years.

Most organisms that lived once are now extinct and, of course, their protein sequences are lost. Thus, the original genetic message of the common ancestor specifying iso-1-cytochrome c, regarded as an input, has many outcomes that nevertheless carry the same specificity. The evolutionary processes can be considered random events along a chain (Cullmann and Labouygues, 1987) that have introduced uncertainty into the original genetic message. This uncertainty is measured by the conditional entropy, in the same manner as the uncertainty of random genetic noise is measured (Section 5.3.2). Because the specificity of the modern iso-1-cytochrome c is preserved, although many substitutions have been accepted, this conditional entropy may be subtracted from the source entropy, $H(x)$, to obtain the information content needed to specify at least one iso-1-cytochrome cc sequence or at least one sequence of any other protein for which a list of functionally equivalent amino acids is available.

Because the sample of iso-1-cytochrome c sequences available is only a tiny fraction of all the organisms that have ever lived or even live today, it is wise to include in the list of known functionally equivalent amino acids those that have similar properties. Some of these amino acids may be found in protein sequences in the future. If they are not included in the estimate of the information content or the complexity, the result will be too small.

The information content of the sequence that determines at least one iso-1-cytochrome c molecule is the sum of the information content of each site. The total information content is a measure of the *complexity* of iso-1-cytochrome c (Section 11.1.2). The final result will be obtained by using Equation 5.12 to calculate the information content of a message that determines at least one among the functionally equivalent amino acids at any site in the iso-1-cytochrome c molecule. I shall apply this to the calculation of the number of protein sequences in iso-1-cytochrome c by the Shannon–McMillan–Breiman Theorem, which will produce the solution to certain problems in molecular biology and genetics.

6.2 A prescription that predicts functionally equivalent amino acids at a given site in protein sequences *revisited*

6.2.1 *The functional equivalence of iso-1-cytochrome c sequences in the electron transfer pathway*

In this section, I shall revisit a prescription (Yockey, 1977a) for predicting functionally equivalent amino acids in homologous protein families, bring it up to date and evaluate its usefulness. I call it a *prescription* or an *Ansatz* because it does not at this time meet the requirements for a *theory* (Section 1.1), according to Sir Karl Popper (1902–94).

6.2.2 *Representation of amino acid functional equivalence in an abstract Euclidean vector space*

The stereographs shown in Figures 6.1, 6.2, 6.3, and 6.4 show the relationship of the amino acids in an abstract Euclidean space of three dimensions (see Mathematical Appendix). An abstract Euclidean space can be established by the use of a set of orthogonal eigenvectors of the matrix of mutation frequencies. Borstnik and Hofacker (1985) introduced a twenty-dimensional Euclidean space of characteristics spanned by eigenvectors of a property preservation matrix closely related to the Dayhoff matrix of mutation frequencies (Dayhoff, 1976). They showed, using maximum entropy analysis, that regarding protein evolution as a random process, three normalized orthogonal eigenvectors establish a three-dimensional flat Euclidean space, in which each amino acid is represented by a point. The position in this space reflects a proper weighting, derived from experiment, of all the relevant properties of the amino acids including the properties mentioned

Figure 6.1. Stereograph showing the sphere enclosing all amino acids functionally equivalent with Ser and Ala. This sphere encloses Gly, Pro, Val, Ile, and Leu from iso-1-cytochrome c sites 27 and 89 as well as phage λ sites 77 and 81 from Reidhaar-Olson and Sauer (1988). See Table 6.7. Stereograph by Clifford A. Pickover, Ph.D. Printed with permission of International Business Machines Corporation.

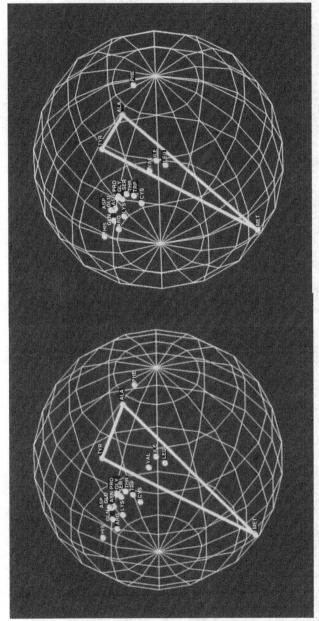

Figure 6.2. Stereograph showing the sphere enclosing all amino acids which are functionally equivalent with Ala, Met, and Tyr. This sphere encloses all amino acids. From sites 86 and 88 of phage λ, Reidhaar-Olson and Sauer (1988). See Table 6.7. Stereograph by Clifford A. Pickover, Ph.D. Printed with permission of International Business Machines Corporation.

61

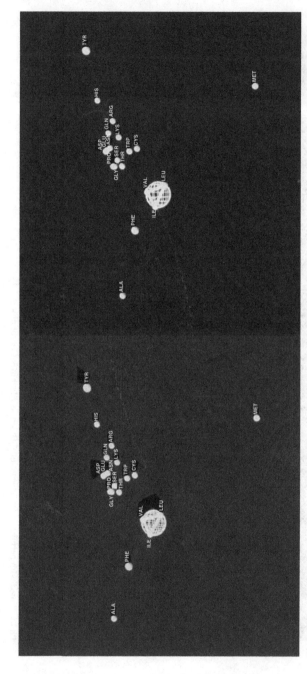

Figure 6.3. Stereograph showing the sphere which encloses Ile, Leu, and Val. No other amino acids are enclosed by this sphere. Ala, Met, Phe, and Tyr are seen at a distance. The rest of the amino acids form a cluster. From iso-1-cytochrome c sites 102 and 103. See Table 6.4. Stereograph by Clifford A. Pickover, Ph.D. Printed with permission of International Business Machines Corporation.

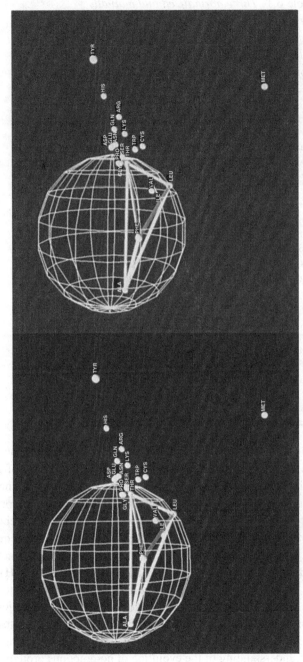

Figure 6.4. Stereograph which shows the sphere enclosing all amino acids functionally equivalent with Ala, Leu, Phe, and Thr. Thr is required to complete Hamming chain to Ala. This sphere encloses Val and Ile. From iso-1-cytochrome c site 43. See Table 6.4. Stereograph by Clifford A. Pickover, Ph.D. Printed with permission of International Business Machines Corporation.

Table 6.1. *Three normalized and mutually perpendicular eigenvectors h_4, h_5 h_6 which are used for the construction of the metric space of polypeptide sequences. From Borstnik and Hofacker (1985) with permission, Adenine Press*

Amino acid	h_4	h_5	h_6
Gly	0.10	0.09	−0.09
Ala	0.08	0.05	−0.90
Pro	0.11	0.10	−0.09
Ser	0.09	0.07	−0.05
Thr	0.08	0.04	−0.09
Gln	0.12	0.13	0.13
Asn	0.10	0.12	0.03
Glu	0.12	0.14	0.02
Asp	0.13	0.15	0.01
Lys	0.11	0.06	0.10
Arg	0.13	0.10	0.21
His	0.08	0.20	0.34
Val	0.03	−0.12	−0.26
Ile	0.02	−0.17	−0.35
Met	0.01	−0.84	0.40
Leu	−0.05	−0.23	−0.22
Cys	0.07	−0.06	0.02
Phe	−0.55	−0.03	−0.45
Tyr	−0.74	0.20	0.42
Trp	0.02	−0.01	0.00

above. By use of the Pythagorean Theorem, we can define the distance between the points that represent the amino acids to as a measure of their relatedness. This statement cannot be made if the vectors that define the space are not orthogonal and therefore mutually independent. The procedures of the prescription can be adapted to a space of any finite number of dimensions but it is very interesting that only three dimensions are adequate. Borstnik, Pumpernik, and Hofacker (1987) continued this work. The method of presentation of the data given by Borstnik and Hofacker (1985) (BH) and by Borstnik et al. (1987), given in Table 6.1, is readily adapted to the prescription. The BH Euclidean eigenvector space and its implementation by the prescription for functional equivalence described below elaborates the rationale of the role played by neutral mutations in the Darwinian paradigm and leads to an understanding of the evolution of homologous proteins and the evolution of *de novo* protein functions (Chapter 12).

The procedure of the prescription for functional equivalence is as follows. One selects all sites with at least two functionally equivalent amino acids in the alignment of the amino acid sequences of a homologous protein family. At each site, the pair is chosen that has the largest BH distance of separation. Consider the sphere that has this distance as its diameter and whose center lies on a line between these two amino acids. The prescription asserts that those amino acids that lie on, or are enclosed by this sphere, are functionally equivalent. If they are not already in the list, they are predicted to be found in the future.

The success of the prescription is a test of the reliability of the set of eigenvectors, reported by Borstink and Hoffacker (1985), to represent the relative functional equivalence of amino acids. If the BH eigenvectors do not adequately reflect the relative functional equivalence the predictions of the prescription will be found to be erroneous. The only erroneous prediction is that of Pro at sites 17 and 41 in iso-1-cytochrome c. The Protein Information Resource (2003) reports inactive iso-1-cytochrome c sequences that contain Pro at these sites.

In order to apply the prescription one needs a table of the distance between all amino acid pairs, the coordinates of the center point between each pair and a means to calculate the radius of each of the twenty amino acids from that center point. According to the Pythagorean Theorem, the distance, D, between any two amino acids whose coordinates are, respectively, (x', y', z') and (x'', y'', z''), is given by Equation 6.1,

$$D = \left[(x' - x'')^2 + (y' - y'')^2 + (z' - z'')^2 \right]^{1/2}. \tag{6.1}$$

The results are the elements of the distance matrix given in Table 6.2 for each pair of amino acids.

The smallest sphere that encloses the region between the two amino acids with this value of D has its center at (x_2, y_2, z_2):

$$x_2 = (x' + x'')/2, \tag{6.2}$$
$$y_2 = (y' + y'')/2, \tag{6.3}$$
$$z_2 = (z' + z'')/2. \tag{6.4}$$

The radius R of all amino acids from the center is calculated from Equation 6.1 substituting the coordinates of each of the twenty amino acids in Table 6.1 for (x', y', z') and (x'', y'', z'') for (x_2, y_2, z_2). The results in the case where Ala-Ser (Table 6.6) determine sphere are shown by the stereo-graph in Figure 6.1.

This sphere must enclose all amino acids known at that site. It sometimes happens that three amino acids are nearly equally distant from each other.

Table 6.2. BH amino acid and hamming distances*

Amino Acid	Trp	Met	Glu	Asp	Lys	Asn	Gln	His	Cys	Tyr	Phe	Ile	Gly	Val	Ala	Thr	Pro	Leu	Arg
Ser	0.12 (1,2)	1.0 (2,3)	0.10 (2,3)	0.11 (2,3)	0.15 (2,3)	0.095 (1,2,3)	0.19 (2,3)	0.41 (2,3)	0.15 (1,2,3)	0.96 (2,3)	0.76 (2,3)	0.39 (1,2,3)	0.05 (1,2,3)	0.29 (2,3)	0.85 (1,2,3)	0.05 (1,2,3)	0.05 (1,2,3)	0.37 (1,2,3)	0.26 (1,2,3)
Arg	0.26 (1,2)	0.97 (1)	0.19 (2,3)	0.21 (2,3)	0.21 (2,3)	0.18 (2,3)	0.086 (2,3)	0.17 (2,3)	0.26 (1,2,3)	0.90 (2,3)	0.96 (2,3)	0.63 (1,2,3)	0.30 (2,3)	0.53 (2,3)	1.11 (2,3)	0.31 (1,2,3)	0.30 (2,3)	0.57 (1,2,3)	
Leu	0.32 (1,2,3)	0.87 (1,2)	0.47 (2,3)	0.48 (2,3)	0.46 (2,3)	0.46 (2,3)	0.53 (2,3)	0.72 (2,3)	0.32 (1,2,3)	1.03 (2,3)	0.59 (1,2)	0.16 (1,2)	0.38 (2,3)	0.14 (1,2)	0.75 (2,3)	0.33 (2,3)	0.39 (2,3)		
Pro	0.17 (2,3)	1.06 (2,3)	0.12 (2,3)	0.11 (2,3)	0.19 (2,3)	0.12 (2,3)	0.22 (2,3)	0.44 (2,3)	0.20 (2,3)	1.00 (2,3)	0.76 (2,3)	0.39 (2,3)	0.01 (2,3)	0.29 (2,3)	0.81 (1,2)	0.07 (1,2)			
Thr	0.12 (2,3)	1.01 (2,3)	0.12 (2,3)	0.16 (2,3)	0.16 (2,3)	0.15 (2,3)	0.24 (2,3)	0.46 (2,3)	0.15 (2,3)	0.98 (2,3)	0.73 (2,3)	0.34 (1,2)	0.05 (2,3)	0.24 (2,3)	0.81 (1,2)				
Ala	0.90 (2,3)	1.58 (2,3)	0.93 (1,2)	0.92 (1,2)	0.92 (1,2)	0.93 (1,2)	1.03 (2,3)	1.25 (2,3)	0.93 (2,3)	1.56 (2,3)	0.78 (2,3)	0.60 (2,3)	0.81 (1,2)	0.66 (1,2)					
Val	0.28 (2,3)	0.98 (2,3)	0.39 (1,2)	0.40 (1,2)	0.41 (2,3)	0.38 (2,3)	0.47 (2,3)	0.68 (2,3)	0.29 (2,3)	1.08 (2,3)	0.62 (1,2)	0.10 (1,2)	0.28 (1,2)						
Gly	0.16 (1,2)	1.06 (2,3)	0.12 (1,2)	0.12 (1,2)	0.19 (2,3)	0.12 (2,3)	0.22 (2,3)	0.44 (2,3)	0.19 (2,3)	0.99 (2,3)	0.75 (2,3)	0.37 (2,3)							
Ile	0.38 (3)	1.00 (2,3)	0.49 (1,2)	0.49 (1,2)	0.51 (2,3)	0.49 (1,2)	0.58 (2,3)	0.79 (2,3)	0.39 (2,3)	1.14 (2,3)	0.60 (1,2)								
Phe	0.73 (2)	1.30 (1)	0.84 (2,3)	0.84 (3)	0.86 (1,2)	0.82 (2,3)	0.90 (2,3)	1.04 (2,3)	0.78 (2,3)	0.92 (1,2)									
Tyr	0.89 (2)	1.28 (2)	0.95 (3)	0.96 (3)	0.92 (3)	0.93 (3)	0.91 (3)	0.82 (2,3)	0.94 (1,2)										
Cys	0.07 (1)	0.87 (3)	0.21 (2)	0.22 (2)	0.15 (2)	0.18 (1,2)	0.23 (2)	0.41 (2)											
His	0.40 (3)	1.04 (3)	0.33 (3)	0.34 (2)	0.28 (3)	0.32 (2)	0.23 (1,2)												
Gln	0.22 (2,3)	1.01 (2,3)	0.11 (1,2)	0.12 (1,2)	0.08 (2)	0.10 (2)													
Asn	0.15 (3)	1.03 (3)	0.03 (1,2)	0.05 (2,3)	0.09 (1)														
Lys	0.15 (2,3)	0.95 (1,2)	0.11 (1,2)	0.13 (2)															
Asp	0.19 (3)	1.07 (1,2)	0.02 (1)																
Glu	0.18 (2,3)	1.06 (3)																	
Met	0.92 (2)																		

* Hamming distances in parentheses

One or two amino acids reported at that site may lie outside the sphere defined by any one of the three pairs. In that case the center of the sphere does not lie on a line between any one of the points in the space. Rather, it lies on a plane determined by the coordinates of the three amino acids. The location of the center of the sphere and its radius can be found by the following procedure:

The general equation of a plane is:

$$x + by + cz = k. \tag{6.5}$$

The values of the parameters, b, c, and k are found by substituting in turn the coordinates of the three amino acids under consideration from Table 6.1 in Equation 6.4 and solving the three linear equations in three unknowns by Cramér's Rule (Mathematical Appendix, Section M1.3). Many programs for solving simultaneous equations are available for personal computers. A sophisticated pocket calculator also has this capability.

The general equation of a sphere is:

$$x^2 + y^2 + z^2 + 2gx + 2fy + 2hz + d = 0. \tag{6.6}$$

This equation may be put in the following form:

$$(x + g)^2 + (y + f)^2 + (z + h)^2 = g^2 + f^2 + h^2 - d. \tag{6.7}$$

The expression on the left of Equation 6.7 is that for the square of the distance, in a flat Euclidean space, from the center of the sphere at $(-g, -f, -h)$ to a point at (x, y, z). The expression on the right of Equation 6.7 is a constant equal to the square of the radius, R, of the sphere:

$$R = [g^2 + f^2 + h^2 - d]^{1/2}. \tag{6.8}$$

The coordinates of the three amino acids will satisfy Equation 6.7 and this provides three linear equations in four unknowns. The fourth equation is found by the condition that the coordinates of the center of the sphere, $(-g, -f, -h)$ must satisfy Equation 6.8. The radius of the sphere and the coordinates of its center are found again by Cramér's Rule. One then proceeds as before to calculate the radius of each residue from the center of the sphere. Two examples of this situation are shown by the stereographs in Figure 6.2 for the case of phage λ sites 86 and 88 (Reidhaar-Olson and Sauer, 1988) and in Figure 6.3 for the iso-1-cytochrome c sites 102 and 103.

It may happen that the method discussed in the preceding paragraph will fail to find a sphere that encloses all the reported residues. This is the case when the four most widely separated amino acids are located on the points of a nearly regular tetrahedron. The center of the sphere does not lie on a plane determined by any three of these amino acids. The solution is found by substituting in Equation 6.5 the coordinates for each point in turn and solving the four linear equations in the unknown parameters f, g, h, d. An example of this situation is shown in the stereograph of Figure 6.4 for the case of iso-1-cytochrome c site 43. The sphere is determined by Ala, Leu, Phe, and Thr.

Referring to Table 6.4, one finds that in iso-1-cytochrome c there are thirteen equivalent amino acids at site 11. In my analysis this site fits nicely among the Ala-Tyr pairs. Because the Ala-Tyr pairs, taken together, contain all the proteineous amino acids except Met, it appears that a sphere is a good approximation to the surface that encloses all functionally equivalent residues. Inspection of the stereographs shown in Figures 6.1, 6.2, 6.3, and 6.4 indicates that the prescription is suitable for its purpose without invoking surfaces more complicated than the sphere.

6.2.3 The prediction of functionally equivalent amino acids in iso-1-cytochrome c

Proteins are very tolerant of amino acids substitutions. The Protein Information Resource (2003) has provided an alignment of iso-1-cytochrome c sites from a number of organisms, together with replacements from other studies, that were found to be either functionally equivalent at that site or not functionally equivalent (<http://pir.georgetown.edu/>). I have used this alignment to apply the prescription to calculate the radius and coordinates of the center of the sphere that encloses all functionally equivalent residues. The amino acids that are found to be on or within the sphere of radius R for each site in the amino acid sequence in iso-1-cytochrome c are shown in Table 6.3. Presenting the results of the calculations in this way makes it easy to correlate the functionally equivalent amino acids with the iso-1-cytochrome c structure.

Inspection of the alignment of the 13 Ala-Gln pairs shows that each of the fifteen predicted amino acids is found at least once except for Trp. In the alignment of ten Ala-Tyr pairs, all the nineteen equivalent amino acids are reported at least once. Similar results are seen in other sets of pairs. Even in the cases where only two pairs are available for comparison, the consistency of the alignment is remarkable.

Table 6.3. *Information content, iso-1-cytochrome c arranged according to sequence order Protein Information Resource (2003)*

Amino acid site	Radius R	Center coordinates (x_2, y_2, z_2)	Functionally equivalent residues known, plain text predicted residues, *italics*	Information content Known residues	Information content All residues
4	0.789	(+0.045, −0.395, −0.25)	(Ala-Met) Gly *Pro* Ser Gln Glu Val Thr Asp Asn *Lys Ile Leu Arg Cys Phe Trp*	0.4147581	1.89160784
5	0.528	(−0.39, +0.075, +0.43)	(Phe-Tyr-Asp-Leu) Gly Pro Ser Thr Gln Asn Glu *Lys Cys Trp*	1.702470	2.00146015
6	0.781	(−0.33, +0.125, −0.24)	(Ala-Tyr) *Gly* Pro Ser Gln Glu Val Thr Asn *Asp* *Lys Ile Leu Arg His Cys PheTrp*	0.120681	2.12320374
7	0.789	(+0.045, −0.395, −0.25)	(Ala-Met) *Gly* Pro Ser Gln Glu Val Thr Asn *Asp* *Lys Arg Ile Leu Cys Phe Trp*	0.4147581	2.49615086
8	0.626	(+0.035, +0.115, −0.279)	(Ala-His-Phe) *Gly* Pro Ser Thr Gln Glu Asp Asp *Asn* *Lys Arg Val Ile Leu Cys Trp*	0.208181	2.57039053
9	0.556	(+0.105, +0.075, −0.355)	(Ala-Arg) Gly Pro Ser *Gln* Glu Val Thr Asn Asp *Lys Ile Leu Cys Trp*	0.737886	2.79186225
10	0.789	(+0.0455, −0.395, −0.25)	(Ala-Met) *Gly* Pro Ser Gln Glu Asp Thr Asn Leu *Ile Lys Arg Val Phe Cys Trp*	0.4147581	1.82603408
11	0.781	(−0.33, +0.125, −0.24)	(Ala-Tyr) Gly Pro Ser *Thr* Asn Glu *Gln* Asp Val *Arg His* Ile Leu Lys Phe *Cys Trp*	0.120681	1.62189866
12	0.781	(−0.33, +0.125, −0.24)	(Ala-Tyr) *Gly* Pro Ser Thr Gln Glu Asn Asp Arg *His Val* Ile Leu Lys Phe *Cys Trp*	0.120681	1.79279439
13	0.556	(+0.105, +0.075, −0.345)	(Ala-Arg) *Gly* Pro Ser Thr *Gln* Asn Glu Glu *Asp* Lys *Val Ile Leu Cys Trp*	0.737886	2.76892422
14			Gly	4.139192	4.139192
15	0.556	(+0.105, +0.075, −0.345)	(Ala-Arg) Gly Pro Ser Thr *Asn* Glu Gln Asp Lys *Val Ile Leu Cys Trp*	0.737868	2.54934943

Table 6.3. (*cont.*)

Amino acid site	Radius R	Center coordinates (x_2, y_2, z_2)	Functionally equivalent residues known, plain text predicted residues, *italics*	Information content	
				Known residues	All residues
16	0.155	(+0.105, +0.07, +0.06)	(Thr-Arg) *Gly Pro Ser Asn Asp Lys Ala Gln Glu Cys Trp*	1.768044	3.26889868
17	0.626	(+0.035, +0.115, −0.279)	(Ala-His-Phe)*Gly* | *Pro* | **Ser Thr Gln Glu Asp Asn Lys Arg Val Ile*Leu *Cys Trp*	0.672312	2.48852781
18	Phe			4.139192	4.13919287
19	0.257	(+0.065, −0.055, −0.125)	(Ile-Lys) *Gly Pro Ser Thr Asn Glu Asp Val Leu Cys Trp*	1.701441	3.12645086
20	0.789	(+0.045, −0.395, −0.25)	(Ala-Met) Gly *Pro Ser Thr Gln Asn Glu Asp Lys Arg Val* Ile Leu Cys Phe Trp	0.4147581	2.40582668
21	0.059	(+0.12, +0.08, +0.155)	(Arg-Lys) Gln	3.914447	3.91444687
22	0.463	(+0.075, −0.005, −0.44)	(Cys-Ala) *Gly Pro Ser Thr Val Ile Leu Tyr (UCU or UCC Ser codons complete Hamming chain.)*	2.144982	4.04886869
23	0.517	(+0.10, +0.09, −0.385)	(Ala-Gln) Gly Pro Ser Thr Asn Glu Asp Lys *Val Ile Leu Cys Trp*	1.183393	3.00489971
24	0.78	(−0.33, +0.125, −0.24)	(Ala-Tyr) *Gly Pro Ser Thr Asn Glu Gln*Asp Lys *Arg His Val Ile Leu Cys Phe Trp*	0.121681	2.80171110
25	Cys			4.139192	4.13919267
26	His			4.139192	4.13919287
27	0.425	(+0.085, +0.06, −0.475)	(Ala-Ser) Gly Pro Thr *Val Ile Leu*	2.438366	3.55642443
28	0.78	(−0.33, +0.125, −0.24)	(Ala-Tyr) *Gly Pro Ser Thr Gln Asn Glu Asp Lys Arg His Val* Ile Leu Phe *Cys*Trp	0.120681	2.54997885
29	0.78	(−0.33, +0.125, −0.24)	(Ala-Tyr) *Gly*Pro Ser Thr Asn Gln Glu*Asp Lys Arg His Val Ile Leu Cys Phe*Trp	0.120681	3.01576131

#		Coordinates	Sequence		
30	0.556	(+0.105, +0.075, −0.345)	(Ala-Arg) Gly Pro Ser Thr Gln Glu Asn *Asp* Val Lys Ile Leu Cys Trp	0.737886	3.17857450
31	0.50	(+0.095, +0.055, −0.40)	(Ala-Lys) Gly Pro Ser Thr Asn Glu Asp *Val* Ile Leu Cys Trp	1.380022	3.54902894
32	0.463	(+0.10, +0.095, −0.44)	(Ala-Glu) Gly Pro Ser Thr Asp *Val Ile Leu Trp (Val required to complete Hamming chain.)*	1.845993	3.40265538
33	0.50	(+0.095, +0.055, −0.40)	(Ala-Lys) Gly Pro Ser Thr Asn Glu *Asp Val* Ile Leu Cys Trp	1.380022	2.85858139
34	0.260	(+0.085, +0.135, +0.145)	(His-Ser) Gln Asn Glu Asp Lys Arg *(Only AGU and AGC codons used for Ser.)*	3.153886	3.87193410
35	0.518	(−0.37, +0.01, +0.12)	(Gly-Leu-Tyr) Ser Thr Lys Gln Trp Asn Glu Cys	2.377720	2.90374348
36	0.517	(+0.10, +0.09, −0.385)	(Ala-Gln) Gly Pro Ses Thr Val Glu Asn Asp Lys Ile Leu Cys Trp	1.183393	3.35303328
37	0.406	(+0.09, +0.07, −0.495)	(Ala-Gly) Pro Thr Val Ile Leu	3.09277087	4.00804533
38			Pro	4.13919287	4.13919287
39	0.466	(+0.09, +0.085, −0.435)	(Ala-Asn) Gly Pro Ser Thr Glu Asp Val Ile Leu Trp *(Thr required to complete Hamming chain.)*	1.76701428	3.69160322
40	Leu			4.139192	4.13919287
41	0.78	(−0.33, +0.125, −0.24)	(Ala-Tyr) Gly |Pro| *Ser Thr Gln Asn Glu Asp Lys Arg His Val Ile Leu Cys Phe Trp	0.468248	2.07804165
42	Gly			4.139192	4.13919287
43	0.461	(−0.121, +0.134, −0.493)	(Ala-Leu-Phe-Thr) Val Ile *(Thr required to complete Hamming chain to Ala.)*	3.060877	3.44335746
44	0.572	(−0.36, +0.015, +0.035)	(Ile-Tyr) Phe Gly Pro Ser Thr Gln Asn Glu Asp Lys Arg His Val Leu Cys Trp	0.468248	3.52899278
45	0.191	(+0.065, +0.00, −0.115)	(Asn-Val) Gly Ser Pro Thr Trp	3.047609	3.63935312
46	Arg			4.139192	4.13919287
47	0.625	(+0.08, +0.125, −0.28)	(Ala-His) Gly Pro Ser Thr Gln Asn Glu Asp Lys Arg Val Ile Leu Cys Trp	0.533821	3.17857450

Table 6.3. (*cont.*)

Amino acid site	Radius R	Center coordinates (x_2, y_2, z_2)	Functionally equivalent residues known, plain text predicted residues, *italics*	Information content	
				Known residues	All residues
48	0.517	(+0.10, +0.09, −0.385)	(Ala-Gln) *Gly Pro Ser Thr Asn Glu Asp Lys Ile Leu Val Cys Trp*	1.183393	3.49085066
49	Gly			4.139192	4.13919287
50	0.12	(+0.10, +0.085, +0.02)	(Gln-Thr) *Gly Pro Ser Asn Glu Asp Lys*	2.744174	3.71635363
51	0.463	(+0.10, +0.095, −0.44)	(Ala-Glu) *Gly Pro Ser Thr Asp Val Ile Leu Trp*	1.845993	3.19202463
52	0.517	(+0.10, +0.09, −0.385)	(Ala-Gln) *Gly Pro Ser Thr Asn Glu Asp Lys Val Ile Leu Cys Trp*	1.183393	2.8017111
53	0.023	(+0.095, +0.08, −0.07)	(Gly-Ser) *(Only AGU and AGC codons used for Ser because of Hamming chain.)*	4.055974	4.05597400
54	0.637	(−0.416, +0.458, −0.064)	(Tyr-Phe-Pro) *Gly Ser Asp Asn Glu (Ser required to complete Hamming chain.)*	3.015761	3.7163563
55	0.50	(+0.095, +0.055, −0.40)	(Ala-Lys)*Gly Pro Glu Ser Thr Asp Asn Val Ile Leu Cys Trp*	1.380002	3.40265530
56	0.46	(−0.32, +0.13, +0.26)	(Tyr-Lys) *Gln Arg His Trp*	3.563888	4.07361910
57	0.025	(+0.85, +0.055, −0.07)	(Thr-Ser)	3.980021	3.98002064
58	0.50	(+0.095, +0.055, −0.40)	(Ala-Lys) *Gly Pro Ser Thr Asn Glu Asp Val Ile Leu Cys Trp*	1.380002	3.24675743
59	0.406	(+0.09, +0.07, −0.495)	(Ala-Gly) *Pro Thr Val Ile Leu*	3.09277087	4.00804533
60	0.525	(+0.041, −0.39, +0.125)	(Asn-Met-Ile) *Ser Thr Lys Arg Val Leu Cys Trp*	2.180741	3.99567012
61	0.556	(+0.105, +0.055, −0.345)	(Ala-Arg) *Gly Pro Ser Thr Gln Asn Glu Asp Lys Val Ile Leu Cys Trp*	0.737886	3.33748421
62	0.556	(+0.105, +0.075, −0.345)	(Ala-Arg) *Gly Pro Ser Thr Gln Asn Glu Asp Lys Val Ile Leu Cys Trp*	0.737886	2.66071471

63	0.789	(+0.045, −0.395, −0.25)	(Ala-Met) Gly Pro Ser Thr Gln Asn Glu Asp Lys Arg Val Ile Cys Phe Trp	0.4147581	3.33232212
64	0.466	(+0.09, +0.085, −0.435)	(Ala-Asn) Gly Pro Ser Thr Glu Asp Val Ile Leu Trp	1.76701428	3.69160322
65	0.517	(+0.10, +0.09, −0.385)	(Ala-Gln) Gly Pro Ser Thr Asn Glu Asp Lys Val Ile Leu Cys Trp	1.183393	3.72360551
66	0.517	(+0.10, +0.09, −0.385)	(Ala-Gln) Gly Pro Ser Thr Asn Glu Asp Lys Val Ile Leu Cys Trp	1.183393	2.21689380
67	0.547	(−0.50, −0.155, +0.08)	(Phe-Tyr-Trp-Leu)	3.835193	3.83519332
68	0.517	(+0.10, +0.09, −0.385)	(Ala-Gln) Gly Pro Ser Thr Asn Glu Asp Lys Val Ile Leu Cys Trp	1.183393	2.74417381
69	0.78	(−0.33, +0.125, −0.24)	(Ala-Tyr) Gly Pro Ser Thr Gln Asn Glu Asp Lys Arg His Val Ile Leu Cys Phe Trp	0.120681	2.84900176
70	0.517	(+0.10, +0.09, −0.385)	(Ala-Gln) Gly Pro Ser Thr Asn Glu Asp Lys Val Ile Leu Cys Trp	1.183393	2.74417376
71	0.34	(+0.055, +0.04, +0.04)	(His-Val) Gly Pro Ser Thr Gln Asn Glu Asp Lys Arg Cys Trp	0.891631	2.70564111
72	0.65	(−0.27, −0.435, −0.025)	(Phe-Met) Leu Gly Ser Thr Val Ile Cys Trp	2.455741	3.95854451
73	0.69	(−0.35, −0.31, +0.15)	(Phe-Tyr-Met-His) Gly Pro Ser Thr Arg Gln Glu Asn Asp Lys Val Ile Leu Cys Trp	0.349184	2.92849389
74	0.78	(−0.33, +0.125, −0.24)	(Ala-Tyr) Gly Pro Ser Glu Gln Thr Phe Asn Asp Lys His Val Ile Leu Arg Cys Trp	0.120681	2.316253541
75	0.46	(−0.645, +0.085, −0.015)	(Tyr-Phe)	3.983295	4.07361910
76	Leu			4.139193	4.13919287
77	0.236	(+0.035, −0.045, −0.10)	(Glu-Leu) Thr Lys Gly Pro Ser Asn Val Cys Trp	2.143615	3.71635363
78	0.024	(+0.115, +0.135, +0.02)	(Asn-Asp) Glu	3.983295	4.07361910
79	0.193	(+0.065, −0.035, −0.22)	(Pro-Ile) Gly Val Ser Thr	3.092770	3.35273380
80	0.076	(+0.10, +0.065, +0.025)	(Arg-Ser) Gln Glu Asp Lys Asn (Only AGU and AGC codons used for Ser.)	3.326916	3.84887310
81	0.057	(+0.115, +0.10, +0.06)	(Lys-Glu) Asn	3.983295	4.07361910

Table 6.3. (*cont.*)

Amino acid site	Radius R	Center coordinates (x_2, y_2, z_2)	Functionally equivalent residues known, plain text predicted residues, *italics*	Information content	
				Known residues	All residues
82	0.543	(−0.363, +0.134, +0.033)	(Phe-Tyr-His-Leu) *Gly Pro Ser Thr* Gln Asn Glu Asp Lys Arg *Cys Trp*	0.949170	3.35970312
83	0.503	(+0.015, −0.505, +0.025)	(Ile-Met) Val Leu (*Cys Trp*) *(Leu needed in the Hamming chain; Cys and Trp are not.)*	3.728728	3.95484676
84	0.195	(+0.03, −0.65, −0.155)	(Pro-Leu) *Gly Ser Thr Val Cys Trp*	2.687381	3.98002064
85	0.096	(+0.105, +0.075,0.005)	(Gly-Lys)*Ser Asn Glu Asp (Only AGU and AGC codons used for Ser.)*	3.533286	3.81896041
86	0.073	(+0.09, +0.08, −0.03)	(Thr-Asn) *Gly Pro Ser*	3.402655	4.04886869
87	Lys			4.139193	4.13919287
88	Met			4.139193	4.13919287
89	0.425	(+0.085, +0.06, −0.475)	(Ala-Ser) *Gly Pro Thr* Val Ile *Leu*	2.438366	3.59456655
90	Phe			4.139192	4.13919287
91	0.414	(+0.165, +0.035, +0.102)	(Ala-Gly-Thr) Val *Ile*	3.167274	3.37790497
92	Gly			4.139192	4.13919287
93	0.650	(−0.27, −0.435, −0.025)	(Phe-Met) *Gly Ser Thr Val Lys Ile Leu Cys Trp*	2.274431	3.78994881
94	0.111	(+0.115, +0.115, +0.02)	(Gln-Pro) *Ser Asn Glu Lys Asp (Gln required to complete Hamming chain between Lys and Pro)*	3.271508	3.71635363
95	0.50	(+0.095, +0.055, −0.04)	(Ala-Lys) *Gly Pro Ser Thr Asn Asp* Val Ile Glu *Leu Cys Trp*	1.380002	3.71635363
96	0.781	(−0.33, +0.125, −0.24)	(Ala-Tyr) *Gly Pro Ser Thr Asn Glu Gln Asp Lys Arg His Val Ile Leu Cys Phe Trp*	0.120681	2.54934943
97	0.517	(+0.10, +0.09, −0.385)	(Ala-Gln) *Gly Pro Ser Thr Asn Glu Asp Lys Val Ile Leu Cys Trp*	1.183393	2.74417381
98	0.061	(+0.125, +0.14, +0.07)	(Gln-Asp) Glu *(Asx identified as Asn.)*	3.876897	3.98329492

Site					
99	0.286	(+0.04, −0.065, −0.005)	(Arg-Leu) Gly Pro Ser Thr Gln Asn Glu Asp Lys Val Cys Trp	1.343515	3.72967730
100	0.517	(+0.10, +0.09, −0.385)	(Ala-Gln) Gly Pro Ser Thr Asn Glu Lys Asp Val Ile Leu Cys Trp	1.183393	2.99101090
101	0.169	(+0.105, +0.175, +0.175)	(His-Asp) Gln Asn Glu Lys	3.630676	3.98329492
102	0.081	(−0.0097, −0.189, −0.277)	(Ile-Leu-Val)	3.813929	3.81392969
103	0.081	(−0.0097, −0.189, −0.277)	(Ile-Leu-Val)	3.813929	3.81392969
104	0.50	(+0.095, +0.055, −0.40)	(Ala-Lys) Gly Pro Ser Thr Asn Asp Val Ile Glu Leu Cys Trp	1.380002	3.63935312
105	0.527	(−0.40, +0.067, +0.044)	(Tyr-Leu-Phe-Gln) Gly Pro Ser Thr Asn Glu Lys Cys Trp	1.891469	3.78657361
106	0.436	(−0.02, −0.535, +0.09)	(Leu-Met)	4.071295	4.07129599
107	0.625	(+0.08, +0.125, −0.28)	(Ala-His) Gly Pro Ser Thr Gln Glu Asp Lys Arg Val Ile Leu Cys Trp	0.533821	2.72628848
108	0.517	(+0.10, +0.09, −0.385)	(Ala-Gln) Gly Pro Ser Thr Asn Glu Asp Lys Val Ile Leu Cys Trp	1.183393	2.62088490
109	0.50	(+0.095, +0.055, −0.40)	(Ala-Lys) Gly Pro Ser Thr Asn Asp Val Ile Glu Ile Leu Cys Trp	1.380002	2.90007176
110	0.50	(+0.095, +0.055, −0.40)	(Ala-Lys) Gly Pro Ser Thr Asn Glu Asp Val Ile Leu Cys Trp	1.380002	3.33708161
111	0.517	(+0.10, +0.09, −0.385)	(Ala-Gln) Gly Pro Ser Thr Asp Asn Glu Lys Val Ile Leu Cys Trp	1.183393	2.84900176
112	0.517	(+0.10, +0.09, −0.385)	(Ala-Gln) Gly Pro Ser Thr Asp Asn Glu Lys Lys Val Ile Leu Cys Trp	1.183393	3.26889868
113	4.139192		Glu	4.139192	4.13919287
			TOTAL	233.18629780	371.417473571
			110 sites Ave.	2.11987543	3.376522487

* Pro is predicted by the prescription but found by not to be functionally equivalent, Protein Information Resource (2003), pirmail@georgetown.edu (March 18, 2003)

Table 6.4. *Values of R from the Phe-Tyr center in sites 54, 75 and 90*

Gly 0.75	Gln 0.78	Arg 0.81	Leu 0.70
Ala 1.14	Asn 0.75	His 0.82	Cys 0.73
Pro 0.76	Glu 0.77	Val 0.75	Phe 0.46
Ser 0.74	Asp 0.78	Ile 0.79	Tyr 0.46
Thr 0.73	Lys 0.76	Met 1.21	Trp 0.67

Another criterion for inclusion in the list of functionally equivalent amino acids is that there must be a path by steps of no more than one Hamming distance connecting all members of the list. The number of posistions in which synomous code words differ is called the Hamming distance (Hamming, 1950). The Hamming distance between each pair of amino acids is given in Table 6.2. The more amino acids enclosed by the sphere the easier it is to find a Hamming chain that includes all candidates. I have inspected each site for assurance that a Hamming chain exists that connects all amino acids. Val is required to complete the chain at sites 32 and 51. Thr is required at sites 39 and 64. The chain between Cys and Ala at site 22 requires the UCU or UCG codons of Ser. Only the AGU and AGC codons of Ser are required at sites 34, 53, 80, and 85. I calculated the information content assuming that the UCN codons of Ser are unassigned so that the total number of assigned codons is 57 rather than 61. At site 83, Leu is needed in the Hamming chain, but Cys and Trp are two Hamming distances from the chain that connects the other amino acids. I have selected the *Gln*-Pro pair at site 94, even though Gln is not reported by the Protein Information Resource (2003) in order to complete the Hamming chain between Lys and Pro. I have used the quadruplet (Ala-Leu-Phe-*Thr*) to determine the sphere at site 43 for the same reason. The sphere determined by (Ala-Met-Tyr) includes all amino acids (Table 6.5) and so the site has zero information content.

The prescription may not be falsified in the usual sense, as it does not purport to predict amino acids that may be reported in the future if they lie *outside* the sphere as it is known at present. It won't be found very useful if a number of amino acids that lie well *inside* the sphere are found not to be functionally equivalent. All the amino acids that are reported to be nonfunctional by the Protein Information Resource (2003) were found to lie outside the sphere as shown in the lists in Table 6.3. The only failure of the prescription is that Pro is predicted at sites 17 and 41, whereas

Table 6.5. *Function equivalent amino acids in iso-1-cytochrome c, phage T4 Lyxozyme[1] and phage λ repressor.[2,3]. Table 6.6 data from Protein Information Resource (2003)*

Source sequence	Site	Radius R	Center coordinates (x_2, y_2, z_2)	Functionally equivalent residues known, plaint text predicted residues, *italics*	Information content	
					All residues	Known
Phage λ	75	0.517	$(+0.10, +0.09, -0.385)$	(Ala-Gln) *Gly Pro Ser Thr Asn* Glu Asp *Lys Val* *Ile Leu Cys Trp*	1.183303	3.33708
Cyto c	23	0.517	$(+0.10, +0.09, -0.385)$	(Ala-Gln) Gly*Pro* Ser*Thr* Asn Glu *Asp Lys Val* *Ile* Leu Cys Trp	1.183393	3.00489971
	36	0.517	$(+0.10, +0.09, -0.385)$	(Ala-Gln) *Gly Pro* Ser Thr Val Glu *Asn Asp Lys* *Ile* Leu Cys Trp	1.183393	3.35303328
	48	0.517	$(+0.10, +0.09, -0.385)$	(Ala-Gln) *Gly Pro* Ser Thr *Asn* Glu *Asp Lys Ile* Leu *Val* Cys Trp	1.183393	3.49085066
	52	0.517	$(+0.10, +0.09, -0.385)$	(Ala-Gln) *Gly* Pro Ser *Thr Asn* Glu Asp Lys Val *Ile Leu Cys Trp*	1.183393	2.8017111
	65	0.517	$(+0.10, +0.09, -0.385)$	(Ala-Gln) *Gly Pro Ser Thr Asn Glu Asp Lys* Val *Ile Leu Cys Trp*	1.183393	3.72360551
	66	0.517	$(+0.10, +0.09, -0.385)$	(Ala-Gln) Gly *Pro Ser Thr* Asn Glu Asp Lys Val *Ile Leu Cys Trp*	1.183393	2.21689380
	68	0.517	$(+0.10, +0.09, -0.385)$	(Ala-Gln) *Gly Pro Ser Thr* Asn Glu Asp Lys Val *Ile Leu Cys Trp*	1.183393	2.74417381
	70	0.517	$(+0.10, +0.09, -0.385)$	(Ala-Gln) Gly Pro Ser *Thr* Asn Glu Asp Lys *Val* *Ile Leu Cys Trp*	1.183393	2.74417376
	97	0.517	$(+0.10, +0.09, -0.385)$	(Ala-Gln) Gly*Pro* Ser Thr Asn Glu Asp Lys *Val* *Ile Leu Cys Trp*	1.183393	2.74417381
	100	0.517	$(+0.10, +0.09, -0.385)$	(Ala-Gln) Gly Pro Ser Thr Asn Glu Asp Lys Lys *Asp* Val *Ile Leu Cys Trp*	1.183393	2.99101090

Table 6.5. (*cont.*)

Source sequence	Site	Radius R	Center coordinates (x_2, y_2, z_2)	Functionally equivalent residues known, plaint text predicted residues, *italics*	Information content All residues	Known
	108	0.517	$(+0.10, +0.09, -0.385)$	(Ala-Gln) *Gly Pro* Ser Thr Asn Glu Asp Lys *Val Ile Leu Cys Trp*	1.183393	2.62088490
	111	0.517	$(+0.10, +0.09, -0.385)$	(Ala-Gln) Gly *Pro* Ser *Thr* Asp Asn Gly Lys *Val Ile Leu Cys Trp*	1.183393	2.84900176
	112	0.517	$(+0.10, +0.09, -0.385)$	(Ala-Gln) *Gly Pro Ser Thr* Asp Asn Glu Lys *Val Ile Leu Cys Trp*	1.183393	3.26889868
Phage λ	77	0.425	$(+0.085, +0.06, -0.475)$	(Ala-Ser) *Gly Pro,* Thr *Val Ile Leu*	2.438366	3.9800
Phage λ	81	0.425	$(+0.085, +0.06, -0.475)$	(Ala-Ser) *Gly Pro Thr Val Ile Leu*	2.438366	3.9800
cyto c	27	0.425	$(+0.085, +0.06, -0.475)$	(Ala-Ser) *Gly Pro Thr Val Ile Leu*	2.438366	3.55642443
	89	0.425	$(+0.085, +0.06, -0.475)$	(Ala-Ser) *Gly Pro Thr* Val Ile Leu	2.438366	3.59456655
Phage λ	83	0.575	$(-0.153, -0.315, +0.155)$	(Met-Gly-Glu-His) Ser Thr Gln Asn Lys Arg *Val Leu Cys*	1.554	1.858
Phage λ	85	0.781	$(-0.33, +0.125, -0.24)$	(Ala-Tyr) *Gly Pro* Ser Thr Gln *Asn* Glu *Asp Lys* Arg*His* Val Ile Leu Cys *Phe* Trp	0.120681	1.733
Cyto-c	6	0.781	$(-0.33, +0.125, -0.24)$	(Ala-Tyr) *Gly Pro* Ser *Gln* Glu Val Thr Asn *Asp* Lys ille Leu *Arg His Cys PheTrp*	0.120681	2.12320374
	11	0.781	$(-0.33, +0.125, -0.24)$	(Ala-Tyr) Gly Pro Ser Thr Asn Glu *Gln* Asp Val *Arg His* Ile Leu Lys Phe *Cys Trp*	0.120681	1.62189866
	12	0.781	$(-0.33, +0.125, -0.24)$	(Ala-Tyr) *Gly Pro* Ser Thr Gln Glu Asn Asp Arg His *Val* Ile Leu Lys Phe *Cys Trp*	0.120681	1.79279439
	24	0.781	$(-0.33, +0.125, -0.24)$	(Ala-Tyr) *Gly Pro* Ser Thr*Asn* Glu Gln *Asp Lys Arg His Val Ile* Leu Cys *Phe Trp*	0.121681	2.80171110
	28	0.781	$(-0.33, +0.125, -0.24)$	(Ala-Tyr) *Gly Pro Ser Thr Asn* Glu *Asp Lys Arg His Val Ile* Leu *Phe Cys Trp*	0.120681	2.54997885

78

			Sequence		
29	0.781	(−0.33, +0.125, −0.24)	(Ala-Tyr) Gly Pro Ser Thr Asn Gln Glu Asp Lys Arg His Val Ile Leu Cys Phe Trp	0.120681	2.84900176
41	0.781	(−0.33, +0.125, −0.24)	(Ala-Tyr) Gly |Pro|*Ser Thr Gln Asn Glu Asp Lys Arg His Val Ile Leu Cys Phe Trp	0.468248	2.07804165
69	0.781	(−0.33, +0.125, −0.24)	(Ala-Tyr) Gly Pro Ser Thr Gln Asn Glu Asp Lys Arg His Val Ile Leu Cys Phe Trp	0.120681	2.31625354
74	0.781	(−0.33, +0.125, −0.24)	(Ala-Tyr) Gly Pro Ser Glu Gln Thr Phe Asn Asp Lys His Val Ile Leu Arg Cys Trp	0.120681	2.31625354
96	0.781	(−0.33, +0.125, −0.24)	(Ala-Tyr) Gly Pro Ser Thr Asn Glu Gln Asp Lys Arg His Val Ile Leu Cys Phe Trp	0.120681	2.54934943
Phage λ 86	0.86	(−0.183, −0.178, −0.114)	(Ala-Met-Tyr) Gly Pro Ser Thr Gln Asn Glu Asp Lys Arg His Val Ile Leu Cys Phe Trp	0.000	1.90384453
Phage λ 88	0.86	(−0.183, −0.178, −0.114)	(Ala-Met-Tyr) Gly Pro Ser Thr Gln Asn Glu Asp Lys Arg His Val Ile Leu Cys Phe Trp	0.00	2.92883644
Phage λ 87	0.436	(−0.02, −0.535, +0.09)	(Leu-Met)	4.071	4.071
Cyto-c 106	0.436	(−0.02, −0.535, +0.09)	(Leu-Met)	4.071	4.071
Phage λ 89	0.789	(+0.045, −0.395, −0.25)	(Ala-Met) Gly Pro Ser Thr Gln Asn Glu Asp Lys Arg Val Ile Leu Cys Phe Trp	0.415	1.599
Cyto-c 4	0.789	(+0.045, −0.395, −0.25)	(Ala-Met) Gly Pro Ser Gln Glu Val Thr Asp Asn Lys Ile Leu Arg Cys Phe Trp	0.4147581	1.89160784
Cyto-c 7	0.789	(+0.045, −0.395, −0.25)	(Ala-Met) Gly Pro Ser Gln Glu Val Trp Asn Asp Lys Ile Leu Cys Phe Trp	0.4147581	2.49615086
Cyto-c 10	0.789	(+0.0455, −0.395, −0.25)	(Ala-Met) Gly Pro Ser Gln GluAsp Thr Leu Ile Lys Arg Val Phe Cys Trp	0.4147581	1.82603408
Cyto-c 20	0.789	(+0.0455, −0.395, −0.25)	(Ala-Met) Gly Pro Ser Thr Gln Asn Glu Asp Lys Arg Val Ile Leu Cys Phe Trp	0.4147581	2.40582668
Cyto-c 63	0.789	(+0.0455, −0.395, −0.25)	(Ala-Met) Gly Pro Ser Thr Gln Asn Glu Asp Lys ArgVal Ile Leu Cys Phe Trp	0.4147581	3.33232212
Phage λ 90	0.790	(+0.047, −0.354, −0.222)	(Ala-Met-His) Gly Pro Ser Thr Gln Asn GluAsp Lys Arg Val Ile Leu Cys Phe Trp	0.208	2.694

Table 6.5. (*cont.*)

Source sequence	Site	Radius R	Center coordinates (x_2, y_2, z_2)	Functionally equivalent residues known, plaint text predicted residues, *italics*	Information content	
					All residues	Known
Lysozyme	86	0.625	(+0.08, +0.125, −0.28)	(Ala-His) Gly Pro Ser Thr Gln Asn Glu Asp Lys Arg *Val Ile Leu Cys Trp*	0.533821	1.778
Cyto-c	47	0.625	(+0.08, +0.125, −0.28)	(Ala-His) Gly Pro Ser Thr GlnAsn Glu Asp Lys Arg *Val Ile Leu Cys Trp*	0.533821	3.17857450
Cyto-c	107	0.625	(+0.08, +0.125, −0.28)	(Ala-His) Gly Pro Ser Thr Gln Asn Glu Asp Lys Arg *Val Ile Leu Cys Trp*	0.533821	2.72628848
Cyto-c	108	0.625	(+0.08, +0.125, −0.28)	(Ala-His) Gly Pro Ser Thr Gln Asn Glu Asp Lys Arg *Val Ile Leu Cys Trp*	0.533821	2.62088490
Phage λ	91	0.463	(+0.075, −0.005, −0.44)	(Ala-Cys) Gly Pro Ser Thr Val Ile Leu Trp *(Gly, Pro, Ser, Thr, Val or UUA, UUG of Leu*	2.145	2.842
Cyto-c	22	0.463	(+0.075, −0.005, −0.44)	(Ala-Cys) Gly Pro Ser Thr Val Ile Leu Trp *(required by Ala-Cys Hamming chain.)*	2.145	4.049
Lysozyme	157	0.626	(+0.035, +0.115, −0.279)	(Ala-Phe-His) Gly Pro Ser Thr Gln Asn Glu Asp Lys Arg *Val Ile Leu Cys Trp*	0.672312	1.388677
Cyto-c	8	0.626	(+0.035, +0.115, −0.279)	(Ala-Phe-His) Gly Pro Ser Thr Gln Glu Asp Asn Lys *Arg Val Ile Leu Cys Trp*	0.208181	2.57039053
Cyto-c	17	0.626	(+0.035, +0.115, −0.279)	(Ala-Phe-His) Gly |Pro| *Ser ThrGln Glu Asp Asn*Lys *Arg Val Ile Leu Cys Trp*	0.672312	2.48852781

* Found by Data from Protein Information Resource (2003) not to be functionally equivalent.

[1] T4 Lysozyme 157 from Alber et al. (1987). *Nature* **330** 41–46, T4 Lyzome 86 from Alber et al. (1988). *Science* **239** 631–5.

[2] Phage k from Reidhaar-Olson & Sauer (1988). *Science* **241** 53–7. Phage λ sites, 75–83 from Bowie et al. (1990).

[3] Phage λ sites 84–91 from Reidhaar-Olson & Sauer (1988).

Hampsey et al. report that iso-1-cytochrome c sequences that contain Pro at these sites are inactive. At the site 34 (my site 37) Das et al. (1989) find that the Gly-Ser mutation renders the protein inactive. This is in accordance with the prescription, as Ser is not within the sphere of functionally equivalent amino acids. One should note that this mutation is unlikely in Nature, because these amino acids are two Hamming steps apart. A mutation Asn-Ile at site 57 (my site 60) restores the function of the protein. At site 38 (my site 41), the mutation His-Pro renders the protein nonfunctional. Function is restored again by an Asn-Ile mutation.

The prescription may be used to test the invariability of a site. For example, the site Phe-87 (Phe-90 my numbering) is phylogenetically invariant. To test this and to investigate the electron transfer between iso-1-cytochrome c and iso-1-cytochrome c peroxidase, Liang et al. (1987) have prepared three mutants at site Phe-87 (Phe-90 my numbering); namely, Tyr, Gly, and Ser. They find that when Tyr or Phe occupy that site, the rate of electron transfer from reduced iso-1-cytochrome c to the zinc cytochrome c peroxidase π-cation radical is 10^4 times greater than when the site is occupied by the mutant Gly or Ser. Inspection of sites 75 and 90 in Table 6.3 shows that the Phe-Tyr pair does not include Gly, Ser, or, indeed, any other amino acids. The Ile-Tyr pair includes Phe at site 44. The other amino acids are accommodated by an increase in R and a considerable shift in the center of the sphere as can be seen from Table 6.3. Thus, the prescription predicts exactly what Liang et al. (1987) have reported, namely, that Tyr is functionally equivalent to Phe at iso-1-cytochrome c site 90 and that this list is complete also for sites 75 and 90. Gardell et al. (1985) found that replacing Tyr at site 248 by site-directed mutagenesis in carboxypepidase A by Phe leaves the catalytic constant toward various peptide and ester substrates unchanged. Noren et al. (1989) have utilized developments in molecular biology to incorporate unnatural amino acids in proteins by site-specific methods. They studied the conserved site Phe[66] in β-lactimase and replaced Phe by Tyr-Ala and the Phe analogues π-ΦPhe, π-NO^2Phe. Activity was preserved if Tyr and the Phe analogues were incorporated but not in the case of Ala. This is predicted and explained by Table 6.5, which shows that Ala, next to Met, is the most distant of all amino acids from the Phe-Tyr center. Thus, the prescription predicts correctly the mutability of the Phe-Tyr pair in several unrelated proteins. The relationship of the Phe-Tyr pair with the other amino acids is shown in the stereographs of Figures 6.1, 6.2, 6.3, and 6.4.

6.2.4 Comparison of iso-1-cytochrome c sequences with other protein sequences

Site-directed mutagenesis has been used to study the functional equivalence of amino acid replacements. The purpose of these studies is to find the consequences of the chemical changes in order to predict the stability and activity of the altered protein. The prescription draws only on the experimentally determined functionally acceptable amino acids and not on the properties of the host protein itself. Accordingly, it is an interesting test to compare the properties of sites in different proteins.

Cupples and Miller (1988) have tested changes in activity due to the substitution of twelve different amino acids at four sites in *E. coli* β-galactosidase at sites His^{464} and Met^3, and 13 at Glu^{461} and Tyr^{503} by site-directed mutagenesis. They find that the substitution of Ser, Gln, Tyr, Lys, Leu, Ala, Cys, Glu, Gly, His, Phe, and Pro at Met^3 and His^{464} retained most of the activity. This result is in accordance with a remark to that effect by Cupples and Miller (1988). The substitution of any of these amino acids for Glu^{461} and Tyr^{503} including Asp and Val reduced the activity by two to four orders of magnitude. Thus, Glu^{461} and Tyr^{503} are invariant.

The Cys amino acids at sites 22 and 25 in iso-1-cytochrome c form a sulfur-sulfur bond associated with the heme pocket. Ala is reported at site 22 and this presents the question of the requirement of a sulfur-sulfur bond for folding and formation of the heme pocket. Reference to the need for Cys-Cys sulfur-sulfur bonds in other proteins is helpful. Bovine pancreatic trypsin inhibitor has three disulphide bonds, Cys^{14}/Cys^{38}, Cys^{30}/Cys^{51}, and Cys^5/Cys^{55}. Marks et al. (1987) have reported removing the Cys^{14}/Cys^{38} disulphide bond in bovine pancreatic trypsin inhibitor and replacing them with Ala or Thr. They found that at physiological temperatures bovine pancreatic trypsin inhibitor can fold without Cys^{14}/Cys^{38}. This supports but does not prove that the iso-1-cytochrome c heme pocket may retain its activity if Ala replaces Cys^{22}. One must remember, however, that the Hamming chain requires either the UCU or UCG codons of Ser at this site.

Alber et al. (1987) have used oligonucleotide-directed mutagenesis to replace Thr^{157} with thirteen other amino acids in phage T4 lysozyme in order to measure the thermodynamic stability of these replacements. These amino acids, namely, Asn, Arg, Asp, Cys, Leu, Arg, Ala, Glu, Val, His, Phe, and Lys, are enclosed by a sphere determined by the Ala-Phe-His triplet. The same complement of amino acids is found at sites 8 and 17 in iso-1-cytochrome cc. Alber et al. (1988) reported the eleven functionally equivalent amino acids at Pro[86] in the same protein. The amino acids are enclosed in a sphere determined by the Ala-His pair. Three comparable sites in iso-1-cytochrome c are shown in Table 6.5. Reidhaar-Olson and Sauer (1988) have reported functionally equivalent amino acids at sites 85 through 91 in the helix 5 region of phage 1 repressor. Bowie et al. (1990), from the same laboratory, reported functionally equivalent amino acids in phage λ repressor sites 75, 77, and 79–83. The amino acids at sites 86 and 88 in phage λ repressor are enclosed in the triplet Arg-Met-Tyr as are those in *E. coli* β-galactosidase at sites His[464] and Met[3]. Table 6.5 shows that all other amino acids are enclosed by the sphere determined by this triplet. The information content of these four sites is zero, because any amino acid may occupy these sites. The amino acids at phage λ site 83 are enclosed by the quadruplet (Met-Gly-Glu-His). Cys and Trp are predicted for site 83.

Reidhaar-Olson and Sauer (1988) point out that several amino acids are underrepresented. For example, Pro is not found and they regard this as evidence that Pro is not a functionally acceptable amino acid at these sites. This point is worth investigating further, as Reidhaar-Olson and Sauer (1988) state that they may not have sequenced a number of candidates large enough to be confident that all candidates have been identified. The Gly-Pro pair has the smallest BH distance and, as Gly is found at sites 85, 86, 89, and 90, it is curious that Pro is not also found. The alignment with iso-1-cytochrome c shows Pro always predicted, but actually found only in iso-1-cytochrome c sites 11, 69, and 96. They are shown aligned with phage λ repressor site 85. Richardson and Richardson (1988) tabulated amino acid preferences from 215 α helices from 45 different globular proteins for 16 different positions relative to the helix ends. They found a substantial preference for Gly over Pro at the helix ends. Pro was found to have the lowest preference of any residue at the helix ends. Gly is the only one of the proteinaceous amino acids that is symmetric. This fact is not included in the BH considerations. It may be that, because of its symmetry, Gly is more easily accepted by the requirements of protein folding. A second reason may be that Pro is a cyclic imino (not an amino) acid, so that it differs from the others in having highly restricted torsion angles and no N-H bond. A third reason for the absence of Pro is that the Gly codons and the Pro codons are separated by two or three Hamming distances. Thus, a single base interchange goes to a non-functional amino acid if Gly and Pro are the only functionally acceptable amino acids at that site.

6.3 The information content or complexity of the iso-1-cytochrome C family

We are now prepared to calculate the information content, or the mutual entropy, for iso-1-cytochrome c in the ideal case using Equation 5.12. The results are shown in Table 6.3 at each site for the case of both the amino acids reported and the amino acids reported plus amino acids predicted by the prescription. However, what I am interested in calculating is the information content or complexity of this fundamental function of the iso-1-cytochrome c homologous family.

I assume that the functionally equivalent amino acids at each site are independent of those at any other site. This is not true if there is a linkage between sites that are near each other in the protein folded condition or which form a complex associated with the active pocket of the protein. Such a linkage implies that, if a mutation to a nonfunctionally equivalent amino acid occurs, the activity may be restored by a mutation at another site in the protein sequence (Section 6.3.3). Such linkages do exist. For example, Brantly, Courtney, and Crystal (1988) report that Lys^{290} and Glu^{342} are contiguous in the folded state of $\alpha 1$-antitrypsin and form a salt bridge, Lys^{290} being charged positively and Glu^{342} negatively. These two amino acids are one Hamming distance apart and so a mutation from either to the other destroys the $+, -$ configuration and results in loss of activity and a high risk of emphysema in the patient. The salt bridge configuration is reformed by a second mutation at either site in the protein chain. In any event, the effect of intersymbol influences on the information content can be accounted for by subtracting the conditional entropy associated with the linkage (Shannon, 1951).

According to the Shannon–McMillan–Breiman Theorem (Section 4.1) the number of iso-1-cytochrome sequences, when the information content is 371.42 bits is:

$$2^{371.42} = 6.43518430225 \times 10^{111}.$$

If one makes this calculation using expression (1), the result is:

$$20^{113} = 1.03845927171 \times 10^{147}.$$

Thus, doing the problem correctly, one finds that the 1-iso-cytochrome sequences are only a very tiny fraction $6.19685707266 \times 10^{-36}$ of the total possible sequences (Section 4.1).

These results will be needed in the discussion of the origin of life scenarios in Chapters 8, 9, and 10. They will provide the knowledge needed to apply the Shannon–McMillan–Breiman Theorem to Eigen's speculations about a "master sequence" and "quasi-species," each one having a "value parameter" or "superiority parameter" (Eigen, 1971, 1992, 2002; Eigen and Schuster, 1977, 1979 1982, Eigen, Winkler-Oswatitsch, and Dress, 1988).

6.4 The explanation of overlapping genes given by information theory and coding theory

6.4.1 The impact of the discovery of overlapping genes

The information content of proteins is also important in explaining the phenomena of overlapping genes. The discovery of overlapping genes caused considerable *Angst* expressed by well-respected commentators in *Science* (Kolata, 1977) and in *Nature* (Szekely, 1977, 1978). Kolata (1977) declared: "Studies of these simple organisms have recently yielded results that shake the foundations of the theories of molecular biology. The hypothesis of nonoverlapping genes is a keystone for many genetic theories." Kolata called for a redefinition of the " . . . very concept of a gene." The genetic code exerts a strong degree of coupling between protein sequences coded in different reading frames. It was thought that this would introduce constraints so severe that two protein sequences of any significant length could not have evolved if their genes overlapped.

However, Nature is often full of surprises that upset such well-established beliefs. Overlapping genes have proved to be a common phenomenon. Although first discovered in viruses, the phenomenon appears in insects and in vertebrates. Shmulevitz et al. (2002) have found overlapping genes in segments of an avian retrovirus and Nelson Bay Retrovirus. Fukuda, Washio, and Tomita (1999) found overlapping genes in *Mycoplasma genitalium* and *Mycoplasma pneumoniae*. There are 162 overlapping gene pairs in the genome of *Mycoplasma genitalium* and 203 overlapping gene pairs in *Mycoplasma pneumoniae* according to the TIGR annotation. Extensive overlapping exists in the mammalian reading frames (Cooper et al., 1998; Kozak, 2001). Klemke et al. (2001) have found two reading frames over 256 codons. Two cases in $\Phi \times 174$ were discovered (Barrell, Air, and Hutchinson, 1976; Sanger et al., 1977; Smith et al., 1977), where genes for proteins of 120 and 151 amino acids each overlap the sequences for longer proteins on a single strand of DNA. In addition, short sections in G4 in which all three reading frames are transcribed were found by Shaw et al. (1978). A similar situation was found in the oconogenic virus SV 40 (Durham, 1978; Fiers et al.; 1978, Reddy et al., 1978) but within an intron Henikoff et al. (1986) have found that the *Gart* locus in *Drosophila melanogaster* contains an entire gene encoding a cuticle protein from the opposite DNA strand Spencer, Gietz, and Hodgetts (1986) found overlapping transcription in the dopa decarboxylase region in *Drosophila*. Jankowski et al. (1986) found two proteins overlapping for thirty-three sites in trout DNA. Williams and Fried (1986) found complementary mRNAs transcribed from opposite strands of the same cellular DNA sequence in the mouse. Adelman et al. (1987) found that the gene in the rat that encodes gonadotropin-releasing hormone (GnRH) from one strand of DNA also transcribes a second gene from the opposite strand, SH, to produce an RNA of undefined function.

6.4.2 A comparison of the genetic message in the DNA of Φ × 174 and in overlapping sites in iso-1-cytochrome c

Figure 6.5 shows part of the genetic messages in the overlapping sequences A protein and B protein of the DNA of Φ × 174 from Figure 4 of Smith et al. (1977). If, in the A protein, the third G for Thr goes through the sequence N = G, T, C, A of the cylinder codons ACN (see Mathematical Appendix), Thr is preserved in the A protein. Smith et al. (1977) reported that the corresponding amino acid codon CGA for Ala; in the B protein, mutates to GTA for Val. Further mutations of GTA, to Leu, CTA and Ile, ATA are compatible with the ACN cylinder codons of Thr. Mutation of the third nucleotide in CAG of the Gln-Tyr, Term series in the A protein causes a Glu, Gln Term Lys series in the B protein as reported by Smith et al. (1977). Moving to the right to where Val Asn, and Thr in the B protein are opposite Leu, Thr, and Leu in the A protein, we note that if the cylinder codons for Val in the B protein mutate through a sequence of all the third nucleotides this causes Leu in the A protein to mutate to Ser, TCA and the Term codons TAA, TGA. If the first nucleotide in the ACA codon of Thr which lies between two Leu codons of the A protein mutates, this produces Ser, TCA, Pro, CCA, and Ala GCA in the A protein. It also produces, pairwise, Ile, ATC, Thr, ACC and Ser, AGC in the B protein. If, in the B protein, the second nucleotide of Thr, ACT mutates, we find Asn, AAT, Ser AGT and Ile, ATT. These mutations in the B protein cause the second Leu in the A protein to go to Ile, ATT, Val, GTT, and Phe, TTT.

A Protein

↑ Thr ↑ *Term*↑   ↑ *Term* ↑ Ala ↑ *Phe* ↑
↑**A C A**↑**T AA**↑   ↑**T A A** ↑**G C A**↑*T* **T T**↑
↑ Thr ↑ *Tyr* ↑ Lys ↑*Term* ↑ Pro ↑ Val ↑
↑**A C C**↑**T A**↑*T***AA G** ↑**TG A** ↑C **C A**↑**G T T**↑
↑ Thr ↑ *Tyr* ↑ Lys ↑ Ser ↑ Ser ↑ Ile ↑
↑**A C G**↑**T AC**↑**AA G** ↑**T** C **A**↑*T* **CA** ↑**A T T**↑
Met↑ Thr ↑ Gln ↑ Lys ↑ Leu ↑ Thr ↑ Leu ↑ Ser ↑ Asp ↑ Ile ↑

Ser
A T G↑**A C G**↑C **A G**↑**A A G**↑**T T A**↑**A C A**↑C T T↑T C G↑G A T↑A T
T↑TCT
A T↓**G A C**↓**G C A**↓**G A A**↓**G T T**↓**A C A**↓C T↓T T C↓G G A↓T A T↓TTC↓T
↓ Asp ↓ Ala ↓ Glu ↓ Val ↓ Asn ↓ Thr ↓ Phe ↓ Gly ↓ Tyr ↓
Phe↓

↓**G T A**↓C **AA**↓**G T** C↓**AT** C↓**A A T**↓
↓ Val ↓ Gln ↓ Val ↓ Ile ↓ Asn ↓
↓C **T A**↓*T* **T AA**↓**G TG**↓**A C C**↓**AG T**↓
↓ Leu ↓*Term*↓ Val   ↓ Thr ↓ Ser ↓
↓**A T A**↓**A AA**↓**G T A**↓**AG C**↓**AT T**↓
↓ Ile ↓ Lys ↓ Val ↓ Ser ↓ Ile ↓

B Protein

Figure 6.5. Overlapping genes are consistent with molecular biology. A and B protein sequences in Φ × 174 from M. Smith et al. Figure 4. The repeated nucleotides are in bold face. The amino acids reported by M. Smith et al. are in plain text. The nucleotides and amino acids produced by mutations are in italics.

All these mutations produce amino acid sets that are found to be functionally equivalent in iso-1-cytochrome c. It is not going too far to say that the functionally equivalent amino acids found at certain sites in iso-1-cytochrome c also are functionally equivalent in A and B protein. This is another illustration of the phenomenon that sites characterized by the same functionally equivalent amino acids (i.e., are enclosed in the same volume in BH space) are found in unrelated protein sequences (Table 6.6).

6.4.3 Overlapping genes are consistent with the genetic code

The explanation, which I proposed (Yockey, 1979) for overlapping genes, was that the redundance of the genetic code and the redundance because of the information content of a protein family would allow, in some cases, the coding of two and perhaps more, genetic messages in one of the three reading frames of single-stranded DNA and in the six reading frames of double-stranded DNA. Pavesi et al. (1997) found that the redundance in

Table 6.6. *Function equivalent amino acids in iso-1-cytochrome c, phage T4 Lyxozyme[1] and phage λ repressor.[2,3]*
Table 6.6 data from Protein Information Resource (2003)

Source sequence	Site	Radius R	Center coordinates (x_2, y_2, z_2)	Functionally equivalent residues known, plaint text predicted residues, *italics*	Information content	
					All residues	Known residues
Phage λ	75	0.517	(+0.10, +0.09, −0.385)	(Ala-Gln) *Gly Pro* Ser Thr *Asn* Glu Asp Lys *Val Ile Leu* Cys *Trp*	1.183303	3.33708
Cyto c	23	0.517	(+0.10, +0.09, −0.385)	(Ala-Gln) *Gly Pro* Ser *Thr* Asn Glu *Asp Lys Val Ile Leu* Cys *Trp*	1.183393	3.00489971
	36	0.517	(+0.10, +0.09, −0.385)	(Ala-Gln) *Gly Pro Ser Thr* Val Glu *Asn Asp Lys Ile Leu* Cys *Trp*	1.183393	3.35303328
	48	0.517	(+0.10, +0.09, −0.385)	(Ala-Gln) *Gly Pro* Ser Thr *Asn Glu Asp Lys Ile Leu Val* Cys *Trp*	1.183393	3.49085066
	52	0.517	(+0.10, +0.09, −0.385)	(Ala-Gln) *Gly Pro* Ser *Thr Asn* Glu Asp Lys Val *Ile Leu* Cys *Trp*	1.183393	2.8017111
	65	0.517	(+0.10, +0.09, −0.385)	(Ala-Gln) *Gly Pro Ser Thr Asn Glu Asp Lys Val Ile Leu* Cys *Trp*	1.183393	3.72360551
	66	0.517	(+0.10, +0.09, −0.385)	(Ala-Gln) *Gly Pro* Ser Thr Asn Glu Asp Lys Val *Ile Leu* Cys *Trp*	1.183393	2.21689380
	68	0.517	(+0.10, +0.09, −0.385)	(Ala-Gln) *Gly Pro Gly* Pro Ser Thr Asn Glu Asp Lys *Val* *Ile Leu* Cys *Trp*	1.183393	2.74417381
	70	0.517	(+0.10, +0.09, −0.385)	(Ala-Gln) *Gly Pro* Ser *Thr* Asn Glu Asp Lys Val *Val* *Ile Leu* Cys *Trp*	1.183393	2.74417376
	97	0.517	(+0.10, +0.09, −0.385)	(Ala-Gln) *Gly Pro* Ser Thr Asn Glu Asp Lys Val *Val* *Ile Leu* Cys *Trp*	1.183393	2.74417381
	100	0.517	(+0.10, +0.09, −0.385)	(Ala-Gln) *Gly Pro* Ser Thr Asn Glu Lys *Asp* Val *Ile Leu* Cys *Trp*	1.183393	2.99101090

				Sequence		
	108	0.517	(+0.10, +0.09, −0.385)	(Ala-Gln) *Gly Pro* Ser Thr Asn Glu Asp Lys *Val Ile Leu Cys Trp*	1.183393	2.62088490
	111	0.517	(+0.10, +0.09, −0.385)	(Ala-Gln) *Gly Pro* Ser *Thr* Asp Asn Glu Lys *Val Ile Leu Cys Trp*	1.183393	2.84900176
	112	0.517	(+0.10, +0.09, −0.385)	(Ala-Gln) *Gly Pro* Ser *Thr Asp Asn* Glu Lys *Val Ile Leu Cys Trp*	1.183393	3.26889868
Phage λ	77	0.425	(+0.085, +0.06, −0.475)	(Ala-Ser) *Gly Pro Thr Val Ile Leu*	2.438366	3.9800
Phage λ	81	0.425	(+0.085, +0.06, −0.475)	(Ala-Ser) *Gly Pro Thr Val Ile Leu*	2.438366	3.9800
cyto c	27	0.425	(+0.085, +0.06, −0.475)	(Ala-Ser) Gly Pro Thr Val Ile Leu	2.438366	3.55642443
	89	0.425	(+0.085, +0.06, −0.475)	(Ala-Ser) *Gly Pro Thr* Val Ile Leu	2.438366	3.59456655
Phage λ	83	0.575	(−0.153, −0.315, +0.155)	(Met-Gly-Glu-His) Ser Thr Gln Asn Lys Arg *Val Leu Cys*	1.554	1.858
Phage λ	85	0.781	(−0.33, +0.125, −0.24)	(Ala-Tyr) Gly Pro Ser Thr Gln *Asn* Glu *Asp Lys* Arg *His* Val Ile Leu Cys Phe Trp	0.120681	1.733
Cyto-c	6	0.781	(−0.33, +0.125.−0.24)	(Ala-Tyr) Gly Pro Ser *Gln* Glu Val Thr *Asn Asp* Lys Ile Leu *Arg His* Cys *Phe*Trp	0.120681	2.12320374
	11	0.781	(−0.33, +0.125, −0.24)	(Ala-Tyr) Gly Pro Ser Thr Asn Glu *Gln* Asp Val *Arg His* Ile Leu Lys Phe Cys Trp	0.120681	1.62189866
	12	0.781	(−0.33, +0.125, −0.24)	(Ala-Tyr) Gly Pro Ser Thr Gln Glu Asn Asp Arg His *Val Ile* Leu Lys Phe Cys Trp	0.120681	1.79279439
	24	0.781	(−0.33, +0.125, −0.24)	(Ala-Tyr) Gly Pro Ser Thr *Asn* Glu Gln *Asp* Lys *Arg His Val Ile* Leu Cys Phe Trp	0.121681	2.80171110
	28	0.781	(−0.33, +0.125, −0.24)	(Ala-Tyr) Gly Pro Ser Thr Gln *Asn* Glu *Asp Lys* Arg *His* Val Ile Leu Phe Cys Trp	0.120681	2.54997885
	29	0.781	(−0.33, +0.125, −0.24)	(Ala-Tyr) Gly Pro Ser*Thr* Asn Gln Glu Asp Lys Arg *His Val Ile* Leu Cys Phe Trp	0.120681	2.84900176
	41	0.781	(−0.33, +0.125, −0.24)	(Ala-Tyr) Gly |*Pro*|* Ser Thr Gln Asn Glu *Glu* Asp Lys Arg His *Val Ile* Leu Cys Phe Trp	0.46248	2.07804165

89

Table 6.6. (*cont.*)

Source sequence	Site	Radius R	Center coordinates (x_2, y_2, z_2)	Functionally equivalent residues known, plaint text predicted residues, *italics*	Information content — All residues	Information content — Known residues
	69	0.781	(−0.33, +0.125, −0.24)	(Ala-Tyr) Gly Pro Ser *Thr* Gln Asn Glu Asp Lys *Arg His Val Ile* Leu Cys *Phe Trp*	0.120681	2.31625354
	74	0.781	(−0.33, +0.125, −0.24)	(Ala-Tyr) Gly Pro Ser Glu Gln *Thr Phe Asn* Asp Lys *His* Val Ile Leu *Arg Cys* Trp	0.120681	2.31625354
	96	0.781	(−0.33, +0.125, −0.24)	(Ala-Tyr) Gly Pro Ser Thr Asn Glu Gln Asp Lys *Arg His* Val Ile Leu Cys *Phe Trp*	0.120681	2.54934943
Phage λ	86	0.86	(−0.183, −0.178, −0.114)	(Ala-Met-Tyr) Gly Pro Ser Thr Gln *Asn* Glu Asp *Lys Arg His* Val Ile Leu Cys Phe Trp	0.000	1.90384453
Phage λ	88	0.86	(−0.183, −0.178, −0.114)	(Ala-Met-Tyr) *Gly* Pro Ser Thr Gln Asn *Glu Asp* Lys Arg *Val Ile* Leu Cys Phe Trp	0.000	2.92883644
Phage λ	87	0.436	(−0.02, −0.535, +0.09)	(Leu-Met)	4.071	4.071
Cyto-c	106	0.436	(−0.02, −0.535, +0.09)	(Leu-Met)	4.071	4.071
Phage λ	89	0.789	(+0.045, −0.395, −0.25)	(Ala-Met) Gly Pro Ser Thr Gln *Asn* Glu Asp Lys Arg *Val* Ile Leu Cys *Phe Trp*	0.415	1.599
Cyto-c	4	0.789	(+0.045, −0.395, −0.25)	(Ala-Met) Gly *Pro* Ser Gln Glu Glu Val Thr Asp Asn *Lys* Ile Leu *Arg Cys Phe Trp*	0.4147581	1.89160784
Cyto-c	7	0.789	(+0.045, −0.395, −0.25)	(Ala-Met) *Gly* Pro Ser Gln Glu Glu *Val Thr* Asn Asp Lys Ile Leu Cys *Phe Trp*	0.4147581	2.49615086
Cyto-c	10	0.789	(+0.0455, −0.395, −0.25)	(Ala-Met) *Gly* Pro Ser Gln Glu Glu Asp Thr Leu Ile *Lys Arg Val* Phe Cys Trp	0.4147581	1.82603408
Cyto-c	20	0.789	(+0.045, −0.395, −0.25)	(Ala-Met) Gly *Pro* Ser Thr Gln Asn Glu Asp *Lys Arg Val Ile Leu* Cys Phe Trp	0.4147581	2.40582668
Cyto-c	63	0.789	(+0.045, −0.395, −0.25)	(Ala-Met) *Gly Pro* Ser Thr Gln Asn Glu Asp *Lys* Arg *Val Ile* Leu Cys *Phe Trp*	0.4147581	3.33232212

Phage λ	90	0.790	(+0.047, −0.354, −0.222)	(Ala-Met-His) Gly *Pro Ser Thr* Gln *Asn Glu Asp* Lys Arg Val *Ile* Leu Cys *Phe* Trp	0.208	2.694
Lysozyme	86	0.625	(+0.08, +0.125, −0.28)	(Ala-His) Gly Pro Ser Thr Gln *Asn* Glu Asp *Lys* Arg *Val* Ile Leu Cys *Trp*	0.533821	1.778
Cyto-c	47	0.625	(+0.08, +0.125, −0.28)	(Ala-His) *Gly Pro Ser* Thr Gln *Asn* Glu *Asp Lys* Arg *Val Ile* Leu Cys *Trp*	0.533821	3.17857450
Cyto-c	107	0.625	(+0.08, +0.125, −0.28)	(Ala-His) *Gly Pro Ser* Thr Gln *Asn* Glu Asp *Lys* Arg *Val Ile* Leu Cys *Trp*	0.533821	2.72628848
Cyto-c	108	0.625	(+0.08, +0.125, −0.28)	(Ala-His) *Gly Pro Ser* Thr Gln *Asn* Glu *Asp Lys* Arg *Val Ile* Leu Cys *Trp*	0.533821	2.62088490
Phage λ	91	0.463	(+0.075, −0.005, −0.44)	(Ala-Cys) *Gly* Pro Ser Thr Val Ile Leu *Trp* (*Gly, Pro, Ser; Thr; Val or UUA, UUG of Leu*	2.145	2.842
Cyto-c	22	0.463	(+0.075, −0.005, −0.44)	(Ala-Cys) Gly Pro Ser Thr Val Ile Leu *Trp(required by Ala-Cys Hamming chan.)*	2.145	4.049
Lysozyme	157	0.626	(+0.035, +0.115, −0.279)	(Ala-Phe-His) Gly Pro Ser Thr Gln Asn Glu Asp *Lys* Arg Val Ile Leu Cys *Trp*	0.672312	1.388677
Cyto-c	8	0.626	(+0.035, +0.115, −0.279)	(Ala-Phe-His) *Gly* Pro Ser Thr Gln Glu Asp Asn Lys *ArgVal Ile Leu Cys Trp*	0.208181	2.57039053
Cyto-c	17	0.626	(+0.035, +0.115, −0.279)	(Ala-Phe-His) Gly |*Pro*|* *Ser* Thr Gln *Glu Asp* Asn Lys *Arg* IleLeu Cys *Trp*	0.672312	2.48852781

* Found by Data from Protein Information Resource (2003) not to be functionally equivalent.

[1] Alber et al. (1988).

[2] Phage λ sites, 75–83 from Bowie et al. (1990).

[3] Phage λ sites 84–91 from Reidhaar-Olson & Sauer (1988).

[1] T4 Lysozyme 157 from Alber et al. (1987). *Nature* **330**, 41–46; T4 Lyzome 86 from Alber et al. (1988). *Science* **239**, 631–635.

[2] Phage λ from Reidhaar-Olson & Sauer (1988). *Science* **241**, 53–57.

the codons for arginine, leucine, and serine favors overlapping genes. The result of the calculations shown in Table 6.3 strenghen this proposal.

Let us now apply what we have learned about the mutual entropy of sequences. As I remarked in Section 6.2.2, the mutual entropy may be used as a measure of the relatedness of two or more sequences regardless of the origin of the sequences. We may calculate the mutual entropy of events x and y in any two reading frames. As I have shown in Table 6.3, sites that have functionally equivalent amino acids have information content lower than 4.139 bits. Accordingly, this amounts to a flexibility in information content shared between the primary and other reading frames, as I have illustrated in Figure 6.1.

Overlapping genes are a manifestation of the use of the full information content of the DNA sequence to record and transcribe genetic messages. This phenomenon shows that the source can drive two or even three channels of transcription with appropriate algorithmic instruction. The study of overlapping genes gives valuable knowledge about the sophistication of genetic algorithms and their source of functional equivalence of amino acids in protein sequences.

The discovery of overlapping genes does not "shake the foundations of molecular biology." It illustrates the need for an understanding of the mathematical foundations of molecular biology. The only redefinition of "the very concept of the gene" needed is to remind ourselves that the gene is the genetic message and not the material DNA or mRNA. It receives a full quantitative explanation from first principles by the discussion given earlier.

7

Evolution of the genetic code and its modern characteristics

I believe no one will be surprised that a large number of the points considered demand a far fuller, more rigorous, and more comprehensive treatment. It seems impossible that full justice should be done to the subject in this way, until there is built up a tradition of mathematical work devoted to biological problems, comparable to the researches upon which a mathematical physicist can draw in the resolution of special difficulties.

Sir Ronald Aylmer Fisher, (1958)

7.1 Early speculations on the evolution of the genetic code

7.1.1 The difficulty of determining the origin of the genetic code

Many papers have been published with titles indicating that their subject is the origin of the genetic code, but actually the content deals only with its evolution. Authors assume that the origin of the genetic code is inevitable once they have created a scenario that provides the components of an informational molecule.

As I have pointed out:

> The calculations presented in this paper show that the origin of a rather accurate genetic code, not necessarily the modern one, is a *pons asinorum* that must be crossed to pass over the abyss that separates crystallography, high polymer chemistry and physics from biology. (Yockey, 1981, 1992)

The paradox is seldom mentioned that enzymes are required to define or generate the reaction network, and the network is required to synthesize the enzymes and their component amino acids. There is no trace in physics or chemistry of the control of chemical reactions by a sequence of any sort or of a code between sequences. Thus, when we make the distinction between the origin of the genetic code and its evolution we find the origin of the genetic code is unknowable (Chapter 11). We are aware that we must take it as following from the axiom of the existence of life (Bohr, 1933). The existence of life is based on the sequence hypothesis and consequently, as Gamow

(Sections 2.1.1 and 2.2.2) proposed, there must be a code between each of several sequences such as those in DNA, mRNA, and protein. Accordingly, in this chapter I shall discuss the evolution of the genetic code, not its origin, which is unknowable.

It was once thought that the genetic code did not evolve because any change would totally scramble the genetic message and would therefore be lethal. However, Barrell, Bankier, and Drouin (1979) showed that mitochondria use some codon assignments that differ from the standard code. In the event that a codon becomes unassigned the organism continues to exist. A new assignment may occur without being an error because of noise (Oba, Andachi, Muto, and Osawa, 1991; Osawa et al., 1987, 1990, 1992).

We have seen in Table 6.3 that the structure of proteins admits the replacement of up to all amino acids at some sites. Because these amino acids do not all have the same codon assignment, this is *de facto* evidence of a nonlethal change of codon. Therefore, the discoveries of the alternate mitochondrial codons, other variations from the standard code, the overlapping genes, and the readthrough tRNAs of the termination codons show that there is flexibility in the genetic code and that it did evolve, at least in the early period before saturation in the second extension.

After examining certain alternatives, I shall present in this chapter a scenario by which the genetic code may have evolved to its present structure. I shall apply several conjectures about protein synthesis in early primitive organisms according to the principles of coding theory discussed in Chapter 5.

7.1.2 Did the genetic code reach its present form by trying many codes and selecting the best?

Freeland et al. (2000) have analyzed the standard genetic code for error minimization. They find the genetic code to be very near or possibly at a global optimum for error minimization. Is there enough time since the beginning of a clement Earth for the modern genetic code to achieve that condition? Did Nature try a number of possible codes and select the best, as the frozen accident scenario proposes (Crick, 1965)? The plausibility of that proposal can be tested by comparing estimates of the number of codes to be tested with the time available for the origin of life.

Let us go to the geological record to find estimates of the time available for the origin of life. The Earth was formed about 4.6×10^9 years ago (Section 8.3.2). It is well established that life existed in abundance at or before 3.8×10^9 years ago (Mojzsis and Harrison, 2002; Schidlowski, 2002). There are 2.52×10^{16} seconds in the 8×10^8 years between these two events. However, during most of these 8×10^8 years, Gaea was busy with other things, such as getting heavily bombarded by meteorites, comets, and other objects, losing her original atmosphere, outgassing and cooling off to form the oceans. The current best estimate of the time between the formation of the oceans and the origin of life is 2×10^8 years or about 6.3×10^{15} seconds.

Having established how much time is available, let us now consider by elementary algebra how to calculate the number of genetic codes to be tested. The codons are formed from the letters of the genetic alphabet by taking all *permutations* of those letters with replacement. The problem is to calculate the number of ways n objects can be arranged, in an unrestricted fashion, without replacing any of them. That is, we need to calculate the number of *permutations* of n objects taken in r_k-tuples without replacement. If we have n objects to be arranged in n permutations without replacement, there are n objects available for the first state. Once the first object is placed there are $(n - 1)$ objects left for the second state. There are $(n - 2)$ states unoccupied for the third object and so on. Therefore, there are $n!$ permutations of n objects in n states. Suppose we consider dividing the states into two subpopulations, one of t states and the other, of course, of $(n - t)$ states. By the argument just given, there are $t!$ permutations in the first population and $(n - t)!$ in the second one. To determine how many arrangements or combinations there are of the total population, we must divide $n!$ by both $t!$ and $(n - t)!$

$$C(n, t) = \frac{n!}{(n - t)! \, t!}. \tag{7.1}$$

By extending this argument we see that the number of arrangements of n obects in k subpopulations of r_k-tuples without replacement is given by:

$$C(n, k) = \frac{n!}{(r_1! \times r_2! \times \cdots r_k!)}. \tag{7.2}$$

Suppose we have a second set of objects, each one of which we wish to identify by a mapping with the r_k-tuples of the subpopulations of the set of the n objects. The correspondence is between each object in the second set of objects and the subpopulations in the first set. In general, there is more than one object in the subpopulations of the first set of objects. This mapping relationship is what is defined as a code in Section 2.2.1. It is now clear that the reason we must calculate the number of combinations without replacement is that a codon would then be assigned to more than one amino acid and that situation would not constitute a code.

We have now grasped the mathematical ideas necessary to calculate the number of genetic codes under certain specified conditions. Suppose the twenty amino acids, and the sixty-four codons were all present at the origin of the genetic code awaiting assignment. Let us calculate the number of genetic codes with the codon-amino acid assignment typical of the modern standard genetic code. Leu, Ser, and Arg are assigned a six-tuple subpopulation of codons each. Ala, Val, Pro, Thr, and Gly are assigned a four-tuple subpopulation of four codons each, and so on, including the three non-sense codons. Thus these r_k-tuple subpopulations may be arranged in 6!, 4!, 3!, and 2! different ways without replacement. Substituting these numbers in Equation 7.2, we have:

$$\frac{(64!)}{(6!)^3(4!)^5(3!)(2!)^9 \times 1 \times 1} = 1.40 \times 10^{70}. \tag{7.3}$$

Any other arrangement in which all but three codons are assigned to at least one of the twenty amino acids also results in a very large number of this order of magnitude. Clearly, this is an implausibly large number of genetic codes from which the modern standard genetic code is presumed to have been selected by evolution in the 6.3×10^{15} seconds of the Earth's early history during which the origin of life events occurred.

The number of codes is reduced considerably if it is possible to start with fewer than twenty sense codons. The idea of Crick (1968), Wong (1976), and Lehman and Jukes (1988) of starting with fewer than twenty amino acids reduces the number of codes enormously, but still leaves an unbelievably large number from which the modern genetic code was to be chose by the slow processes of natural selection. In addition, as the vocabulary is increased the number of genetic codes increases dramatically as can be seen by substituting the appropriate numbers in Equation 7.2. One must presume that the modern genetic code did not originate from among 1.40×10^{70} codes awaiting assignment.

7.2 Did the genetic code evolve from a first extension of a four-letter alphabet?

7.2.1 Was the genetic code ever binary?

The possibility that the genetic code began with a binary alphabet must be considered first before proposing a four-letter alphabet, because a binary alphabet is the simplest one. Computers and electronic communications use binary alphabets exclusively. The only hint that the genetic code may have begun with a binary alphabet is that the compounds that form the letters of the DNA and RNA alphabet are chemically of two kinds, namely, purines A and G, and pyrimidines, C, T, and U.

There are a number of objections to the belief that the genetic code was binary at some time in its early history. A binary alphabet might have been made from one purine and one pyrimidine. Those who believe in a prebiotic soup (Chapter 8) are aware of the difficulties in the prebiotic formation of pyrimidines. For example, cytosine is obtained in a yield of about 5 percent in an aqueous solution of 1.0 M potassium cyanate and 0.1 M cyanoacetylene held at 100°C for twenty-four hours. This is hardly a reasonable prebiotic synthesis because it is obvious that such controlled conditions require a *deus ex machina* in the form of an expert biochemist.

7.2.2 Jukes' proposal that the genetic code evolved from a doublet code

Jukes (1965, 1966, 1973, 1974, 1981, 1983a, 1983b, 1986, 1987, 1993) first suggested that the current standard triplet code evolved from a doublet code with a four-letter alphabet. Crick (1968) concurred, provided that only the first two nucleotides are read and the third letter is effectively a spacer. He believed this to be necessary because the size of the codon is dictated by the diameter of the double helix. Let us accept the condition that the third nucleotide was a spacer and apply the principles of coding and information theory to examine a scenario for the evolution of the genetic code from a first extension of a quaternary alphabet.

The number of amino acids that can be assigned code words by any code based on the first extension of a four-letter alphabet is no more than fifteen, plus a non-sense or termination codon. So the first test of Jukes' proposal is: Could highly specific proteins have been composed of fewer than twenty amino acids? It is a significant support of this scenario that all twenty amino acids do not appear in some modern proteins. The ferredoxins, for example, all have fewer than twenty kinds of amino acid. As Lehman and Jukes (1988) have pointed out, there are only thirteen amino acids in the clostridial ferredoxins of *Clostridium butyricum*. The seven missing ones are, namely, Arg, Leu, His, Lys, Met, Tyr, and Trp (Dayhoff and Eck, 1978; George et al., 1985). They also pointed out that a comparison of nine different clostridial type ferredoxins shows that there are only six different amino acids that occupy the invariant sites. As I pointed out (Yockey, 1977b), this is consistent with the fact that the cytochrome c message can be written with fourteen amino acids, as inspection of Table 6.4 shows. We may therefore retain for consideration the idea that the first code had fewer than twenty sense codons (Crick, 1968; Jukes, 1965, 1966, 1973, 1974, 1983a, 1983b; Orgel, 1968; Wong, 1976).

It is reasonable that the direction of evolution was toward organisms with genetic codes that had the largest vocabularies and the least number of mutations to non-sense codons. Let us now consider the scenario that a number of independent origin of life events occurred (Raup and Valentine, 1983), each with its own code of a given vocabulary size τ with γ mutations to non-sense codons. Only those mutations at a Hamming distance one are considered because the mutations of two and three Hamming distances are of second and third order and are therefore very improbable.

The following argument is substantially due to Figureau and Labouygues (1981), Cullmann (1981), Labouygues and Figureau (1982, 1984), and Cullmann and Labouygues (1983). The code C is composed of sense code words that have a specific assignment. Other doublets of the alphabet are not part of code C. Let K_i be the number of times that symbol i is used in the first position of code words in the code, and K_j is the number of times the symbol j is used in the second position. We must first calculate the number of codes that have γ mutations to non-sense codons. In general, the number of exchanges in code C of q symbols K_i times in the first position is $q K_i$. We must subtract the number of times symbol i replaces itself, so that the total for symbol i is $(q K_i - K_i^2)$. The same is true for the number of exchanges of q symbols K_j times in the second position. These two expressions are added and a summation is made over only the sense code words in code C. That is, $i \in C$ and $j \in C$.[*]

$$\gamma = \sum_{i \in C}(q - K_i)K_i + \sum_{j \in C}(q - K_j)K_j. \tag{7.3}$$

[*] This is read, "The element i is a member of code C."

The resistance of mutations to non-sense code words in code C is measured by the number of code words, D_1, in code C, that are separated by a Hamming distance of one. The number of pairs of symbols of sense code words and non-sense code words is the square of the number of times the symbol i appears in the first position. The first symbol in the is not one Hamming distance from itself, so the number of pairs of symbol i in the first position must be subtracted from K_i^2 so we have $(K_i^2 - K_i)$. By the same token, the number of pairs of letter j in the second position is $(K_j^2 - K_j)$. These expressions must be added and summed over all i and all j. We have counted each pair twice and so we must multiply by $\frac{1}{2}$.

$$D_1 = \frac{1}{2}\left[\sum_i \in C\left(K_i^2 - K_i\right) + \sum_j \in C\left(K_j^2 - K_j\right)\right]. \qquad (7.4)$$

All other codes of the same size, τ, can be occupied by removing code word ij and replacing it with code word hl. The variations ΔD_1 and $D\gamma$ can be obtained by the following equations, where in each case, $i, j, k \in C$:

$$\Delta D_1 = -\{(K_i - 1) + (K_j - 1) - (K_h + K_l)\} \qquad (7.5)$$
$$\Delta\gamma = 2\{(K_i - 1) + (K_j - 1) - (K_h + K_l)\}. \qquad (7.6)$$

This is easily understood when we see that the removal of code word ij transforms K_i to $(K_i - 1)$, K_j to $(K_j - 1)$. The introduction of doublet hl changes K_h to $(K_h + 1)$ and K_l to $(K_l + 1)$.

The number of doublet codes for each value of γ and τ is calculated from the number of arrangements of the rows and columns of the 4×4 square array of the sixteen boxes in which code words may be found (Cullmann 1981). As we saw in Equation 7.1, the total number of codes for each value of t is given by the number of arrangements of q^2 things taken τ at a time:

$$C(q^2, \tau) = \frac{q^2!}{(q^2 - \tau)!\tau!}, \qquad (7.7)$$

where q is the number of letters in the alphabet, in this case $q = 4$. These calculations have been carried out by Cullmann and Labouygues (1983) as described by Cullmann (1981). The number of doublet codes for each value of γ and τ is shown in Table 7.1.

The total number of all codes for each value of τ is calculated from Equation 7.7 and is shown in the bottom row of Table 7.1. For example, the total number of codes for eight amino acids is 12,870. This is still a large number, although it is much more manageable than the numbers

Table 7.1. *The number of doublet codes as a function of τ and γ*

t	4	5	6	7	8	9	10	11	12	13	14	15	16
γ													
0													1
													1
6												16	
												1	
10											48		
										1			
12	8								8	32	72		
	1								1	1	1		
14	0								0	144			
	0								0	1			
16	324	96			12			96	324	288			
	2	1			1			1	2	1			
18	384	432	288	112	0	112	288	432	384	96			
	2	2	3	2	0	2	3	2	2	1			
20	792	864	1008	864	1008	864	1008	864	792				
	3	2	3	2	4	2	3	2	3				
22	288	1392	1152	1728	1344	1728	1152	1392	288				
	1	4	2	4	4	4	2	4	1				
24	24	1152	2752	2304	3168	2304	2752	1152	24				
	1	2	7	5	6	5	7	2	1				
26		432	1584	3024	2304	3024	1584	432					
		2	4	6	5	6	4	2					
28			1224	2592	3792	2592	1224						
			4	4	8	4	4						
30				816	1152	816							
				1	2	1							
32					90								
					2								
	1820	4368	8008	11440	12870	11440	8008	4368	1820	560	120	16	1

The sum of the large print numbers in each column is the number of combinations of sixteen items taken t at a time. These numbers are given in this row. The small print numbers below each number in large print is the number of different configurations generating those codes. This table is from Cullmann and Labouygues (1983) with permission.

discussed in Section 7.1.3. However, the number of codes that need to be considered can be reduced still more. In the case where $\gamma = 20$ and $\tau = 8$, there are 1,008 codes. One sees immediately that there is a most substantial reduction in the number of doublet codes over the number of triplet codes. Jukes' suggestion passes the second test, which is that the multiplicity of codes in the first extension of a quaternary alphabet is very much smaller than the number of codes found in Section 7.1.3.

7.2.4 Evolution of the genetic code by random walk between Markov states

The third test of Jukes' suggestion is to show how these doublet codes may have evolved by random walk from ones with a vocabulary of, say, six to nine amino acids to the second extension standard triplet modern genetic code with a vocabulary of twenty amino acids and to the other modern genetic codes that differ from the standard genetic code (see random walk in the Mathematical Appendix). It is too dark to see all the way back to the *pons asinorum*, but, to take a specific example, let us consider the genetic code at a time when it had a vocabulary of τ amino acids. Perhaps these were the ones, appearing in the first column of Table 7.2, which, in the second extension, have more than one codons for each amino acid. (I have called them *cylinder codons* for the reasons discussed in the Mathematical Appendix.) In the modern standard genetic code, the third position in these triplet codons has no discriminatory function. The tRNAs recognize all four *cylinder codons*.

In this scenario we presume several different independent origin of life events, each with a doublet genetic code characterized by a pair of numbers (τ, γ). Let the pairs of numbers (τ, γ) in Table 7.1 be a set of Markov states. Mutations will occur that change one code word to another. The mechanism for this is the AT to GC pressure or GC to AT pressure driven by the functional importance or requirements of the protein specificity as recorded in the genome, a reduction in the mutations to non-sense codons and the increase in vocabulary (Jukes et al.,1987; Jukes and Bushan, 1986; Osawa et al., 1987). The progress by random walk to a saturated code, which has a vocabulary of fifteen amino acids and one non-sense codon, can be regarded as a stationary Markov process in which a step to a second Markov state results in a decrease in γ or an increase in τ (Cullmann and Labouygues, 1983).

The minimization of genetic noise has been suggested as a controlling influence in the evolution of the genetic code by numerous authors. To minimize the effect of genetic noise, D_1 will progress to a maximum value and γ to a minimum value. To illustrate the process by which a Markov chain evolves to a minimum in γ and a maximum in D_1, suppose the initial Markov states are the four codons {UU, AA, CC, GG}. The value of γ is twenty-four and the value of D_1 is zero. If one removes the code word

Table 7.2. *A proposed first extension and evolution to the second extension of the genetic code*

First extension

Amino acid	Doublet codon	Triplet codon	mtDNA codon
Gly	GG	GGN	GGN
Pro	CC	CCN	CCN
Leu	CU	CUN**	CUN*
Arg	CG	CGN	CGN
Thr	AC	ACN	ACN
Val	GU	GUN	GUN
Ala	GC	GCN	GCN
Ser	UC	UCN	UCN

First extension

Amino acid	Doublet codon	Triplet codon	mtDNA codon
Phe	UU	UUU, UUC	UUU, UUC
Cys	UG	UGU, UGC	UGU, UGC
Gln	CA	CAA, CAG	CAA, CAG
Asn	AA	AAU, AAC	AAU, AAC
Glu	GA	GAA, GAG	GAA, GAG
Ile	AU	AUU, AUC, AUA	AUU, AUC
Non-sense	UA	UAA^, UAG^	UAA, UAG
Non-sense	AG		

Second extension

Amino acid	Triplet codon	mtDNA codon
Leu	UUA, UUG	UUA, UUG
Trp	UGG	UGG
Non-sense	UGA++	UGA#
His	CAU, CAC	CAU, CAC
Lys	AAA, AAG	AAA, AAG
Asp	GAU, GAC	GAU, GAC
Met	AUG	AUG+, AUA+, AUU+, AUC+
Tyr	UAU, UAC	UAU, UAC
Arg	AGA, AGG	AGA##, AGG##
Ser	AGU, AGC	AGU, AGC

* Codes for Thr in yeast mtDNA. Li and Tzagoloff (1979), Bonitz et al. (1980), Sibler et al. (1981), and Neurospora mtDNA, Heckman et al. (1980). # UGA codes for Trp in animal, fungal, yeast, and protozoa mtDNA, Anderson et al. (1982), Watson et al. (1987). + Met code internal is AUG, AUA; for initiation AUG, AUA, AUU, AUC, Barrell et al. (1980), Montoya et al. (1980), Anderson et al. (1981), Bibb et al. (1981). ^ Codes for Gln in *Paramecium*, Klug and Cummings (1986) in *Tetrahymena*, Horowitz and Gorovsky (1985), Kuchino et al. (1985). ## Non-sense in mammalian mtDNA, Barrell et al. (1980), Anderson et al. (1981), Watson et al. (1987). ** CUG codes for Ser in yeast, Kawaguchi et al. (1989). ++ Codes for Trp in *Mycoplasma caprolium*, Yamao et al. (1985), Muto et al. (1986), and yeast Macino et al. (1979), Codes for selenocysteine, Leinfelder et al. (1988).

GG and substitutes UG then, $\gamma = 22$ and $D_1 = 1$. If the Markov chain proceeds in this way to the evolution of code $\{UU, UC, UA, UG\}$, then $\gamma = 12$ and $D_1 = 6$. If τ is increased to six with the additional assignments to UC and CC, $\gamma = 10$ and $D_1 = 9$. If the step to a second Markov state is accomplished by a decrease in γ, the vulnerability of the genetic message to genetic noise is reduced.

7.2.5 Expansion of the genetic vocabulary and decreasing the number of possible genetic codes

As this process of evolution proceeds by random walk toward the Markov states to the right side of Table 7.1, the number of codes available to the survivors of this evolutionary process decreases naturally, for mathematical reasons. Origin of life events that start at any Markov state in Table 7.1 evolve up and to the right. All possible doublet codes that may have originated in any Markov state (τ, γ) will coalesce into one of the smaller number of possible Markov states as the code word vocabulary increases. In the case of the fourteen sense doublet codons where $\gamma = 10$, there are only forty-eight codes available. This is reduced to sixteen codes as the fifteenth amino acid is admitted. If more than one organism had followed a random walk, the different codes of those organisms would have been squeezed through a bottleneck upon arriving at fourteen sense codons and tended to assume similar but not necessarily identical genetic codes. It is at this stage in the scenario that the transition to the second extension must occur. In support of this scenario the number of known mitochondrial codes together with other genetic codes that differ from the standard code and the standard code itself is nearly the same as the number of codes allowed in the last two Markov states in Table 7.1.

It is important to note that this scenario does not require that a substantial fraction of the available codes be tested, and, therefore, Jukes' suggestion passes the third test. The evolutionary process follows an ascending path, so to speak, to the doublet code saturation bottleneck at fourteen to fifteen sense code words.

This process is irreversible, as all backward steps increase the vulnerability to mutations to non-sense codons. For the same reason the paradigm assumes that once an assignment is made it does not change. Each code evolves naturally in this way to a larger vocabulary giving it greater protection from mutations to non-sense code words. Furthermore, after emergence from the bottleneck, the situation reverses and then codon sense reassignments are indistinguishable from genetic noise, as Crick (1968) and Orgel (1968) have pointed out. For these reasons, one code cannot evolve from another. The only exception is when a bottleneck is approached so that the number of codes decreases and new codon assignments do not create genetic noise.

7.3 Proposed doublet codon assignments and evolution to the modern genetic codes

7.3.1 The formation of triplet codons from doublet codons

The next step in the evaluation of Jukes' proposal is to consider the actual codon assignments as they are found in the modern standard code and in the several mitochondrial codes. In Yockey (1977b, Table 4), following Jukes' work, I made a proposal for codon assignments, which is given here with several modifications.

The proposed codon assignments in the first and second extension of the quaternary alphabet are shown in Table 7.2. Wong (1988) suggested that Gly, Ala, Ser, Asp, Glu, Val, Pro, Thr, Leu, and Ile were the first amino acids to be assigned codons. I have assumed that the amino acids in the left column in Table 7.2 were the first amino acids to be assigned codons. They have cylinder codons in the second extension of the genetic code (see cylinder codons in the Mathematical Appendix). Six additional amino acids were assigned doublet codons later in the first extension as shown in the second column of Table 7.2. The doublet UA became the stop codons UAA and UAG in the second extension. As the second extension emerged, the remaining amino acids of the modern twenty were assigned the codons as shown in the third column of Table 7.2. In the second extension, Cys retained UGU and UGC and Trp was assigned UGG. UGA is a non-sense codon but codes for Trp in *Mycoplasma caprolium* (Macino et al., 1979; Yamao et al., 1985), in *Neurospora crassa* (Heckman et al., 1980), and in yeast (Macino et al., 1979). UAG and UAA are non-sense in the first extension but code for Gln in *Paramecium* (Klug and Cummings, 1986) and Glu in *Tetrahymna* (Horowitz and Gorovsky, 1985). UAG codes 1–2 percent for Glu in mouse cells infected with Moloney murine leukemia virus (Kuchino et al., 1987). Gln retains CAA and CAG, whereas His is assigned CAU and CAC. Asn retains AAU and AAC and Lys is assigned AAA and AAG in the second extension. Arg and Ser divide between them the previously non-sense codons, AGN. This must have been somewhat arbitrary, as in

Drosophila yakuba mtDNA AGA is assigned to Ser rather than Arg (Clary and Wolstenholme, 1985; Fox, 1987).

This scenario is supported by the following facts: (a) the AGU and AGC are separated by two or three Hamming distances from the UCN cylinder codons of Ser: and (b) the AGA and AGG are non-sense codons in mammalian mtDNA (Barrell et al., 1980; Anderson et al., 1981). That is, the AGA and AGG codons were not assigned to amino acids in the mtDNA mammalian genetic code. The assignment or nonassignment of the AGA and AGG codons also supports the scenario that they were assigned at the time of the second extension and that Arg and Ser were included in the original set of amino acids with cylinder codons. That is, those with more than one codon assigned to an amino acid.

This paradigm explains the puzzling fact that Arg, which is relatively rare compared to Lys in modern proteins, has six codons, whereas Lys has only two. Lys is one Hamming distance from Asn and the BH distance is 0.09. The assignment requires only the establishment of specificity in the third nucleotide typical of the process of going to the second extension. By contrast, the Hamming distance between the CGN cylinder codons of Arg and the AAA, AAG codons of Lys is 2.3. The BH distance between Arg and Lys is 0.12. Jukes (1983a) suggested that Lys was a late entry to the code.

The AGA and AGG codons and the AGU and AGC codons were available non-sense codons and became assigned to Arg and to Ser, respectively, in a manner that parallels the *Stop Codon Takeover Model* of Lehman and Jukes (1988).

> The basic premise of that model is that all message ribotrinu-
> cleotides were effectively 'stop' (protein chain terminating) codons
> before adaptor molecules, via their anticodon bases, could evolve
> to bind tightly to them. In contrast, previous models propose that
> at any given point in time most or all codons specified an amino
> acid and that code evolution proceeded as newer amino acids some-
> how acquired codons from older amino acids. (Lehman and Jukes,
> 1988)

Additional support of the assignments in Table 7.2 comes from the empirical fact that Lys, Tyr, His, Met, and Trp are missing or very rare in ferredoxins and are, therefore, perhaps, the last to be incorporated in the genetic code. These amino acids are all in the column headed *Second Extension* in Table 7.2. Asp is the only one of those amino acids incorporated in ferredoxin from *Clostridium thermoaceticum*, which is located only in the third column of Table 7.2. The ferredoxins are often regarded as very ancient proteins that were present in ancestral organisms soon after the origin of life (George et al., 1985). They are iron-sulfur electron carriers and may be related to the interesting suggestions of Wöchtershöuser (1988a, 1988b, 1988c, 1990, 1997, 1998, 2000) regarding the role of FeS in the origin of life. Of the amino acid pairs assigned to the first and second extensions in Table 7.2, only Asp and Glu could be exchanged without violating the need for observing the biosynthetic pathways. Met and Trp must have received their single codons after the second extension was well along. The twelve amino acids admitted in saturating the first and second extensions could have been added very smoothly to the vocabulary. Once the saturation of the second extension triplet genetic codes was established, there was little if any possibility for further changes because unless a codon becomes unassigned and available for reassignment, changes are indistinguishable from genetic noise (Böck, 2002; Castresana, 1998; Crick, 1968; Oba et al., 1991; Orgel, 1968).

7.3.2 mRNA editing in the mitochondrial codes

CGG was reported to be a Trp codon in *Zea maize* mtDNA by Fox and Leaver (1981) and in the mtDNA of *Oenothera berterilana* by Hiesel and Brennicke (1983). No tRNA specific for Trp and recognizing CGG has been found (Maréchal et al., 1985). Although the CGG codons exists in the wheat mtDNA, Gualberto et al. (1989) found that the codon UGG is at the corresponding site in the mRNA sequence so that the actual codon read is UGG, the standard codon for Trp. Covello and Gray (1989) found only C →U editing, thus the plant mitochondrion code does not differ from the standard genetic code. Other C →U conversions in the first and second position were found by Gualberto et al. (1989), which edited His → Tyr, Ser → Leu, Ser → Phe, Leu → Phe, and Pro → Leu. Covello and Gray (1989) also found C → U editing in maize, rice, wheat, pea, soybean, and *Oenethera*. Hiesel et al. (1989) find that the Arg codon CGG is often edited to the Trp codon UGG in higher plant mitochondria. This suggests that the standard genetic code is used in plant mitochondria and this resolves the frequent coincidence of CGG codons and Trp in different plant species.

mRNA editing changes a Gln CAA codon to UAA at site 2152 in apolipotein B mRNA, which allows the translation of two proteins of different length from the same gene (Powell et al., 1987; Shaw et al., 1988; Tennyson et al.,

1989a; Tennyson et al., 1989b), according to whether or not the editing event occurs. Simpson and Shaw (1989) have reviewed mRNA editing in mitochondria.

Section 7.3.3   *The biosynthetic pathways*

The amino acids assigned cylinder codons (see the Mathematical Appendix) in the first column of Table 7.2 are those that are not formed in biosynthetic pathways from other amino acids. The assignment of codons to amino acids assigned codons in the second and third columns was made on the ground of the analysis by Wong (1975, 1976, 1981, 1988, 2002) of the biosynthetic pathways of formation. The six amino acids admitted in the second extension (Table 7.2) have biosynthetic pathways requiring amino acids that were assigned codons in the first extension. The biochemistry must be involved in the coevolution of the genetic code and the amino acids that are added to increase the vocabulary by the increase of τ.

This scenario shares much with Wong's coevolution theory, because it also contemplates the genetic code expanding its vocabulary together with the evolution of the amino acids. It also shares with the Stop Codon Takeover Model of Lehman and Jukes (1988) the idea that the evolution of the genetic code must be such that mutations to non-sense codons is minimized. An advantage of this scenario is that it follows the mathematical theory of coding that applies to all codes and thereby avoids *ad hoc* reasoning. Furthermore, it places the burden of evolution of the genetic code on satisfying the specificity needs of the evolving proteins in the primitive organisms rather than, as Weber and Miller (1981) have suggested, on a Procrustean bed of the abundance of amino acids in a phantom *Urschleim* (Section 8.3).

7.4 Characteristics of the genetic code

7.4.1 *The genetic code is instantaneous*

In the course of the evolution of the genetic code, as the genetic code approaches and emerges from the bottleneck that leads to the modern second extension code, specificity must be assigned to the third nucleotide of each codon. The code must be instantaneously decodeable at all times in its evolution in order to avoid the confusion that would occur, for example, if the genetic code had the doublet AA for Asn together with the triplet code AAA for Lys. Therefore, the doublet codons in the second column in Table 7.2

must assign specificity, one by one, to the third nucleotide of codons that are assigned to amino acids that enlarge the vocabulary. However, it is not necessary for additional specificity to be assigned to all doublets. For example, the cylinder codons (see the Mathematical Appendix) in the first column on Table 7.2 continue to function as doublet codons.

As the transition to the second extension proceeds, only those newly admitted amino acids must be assigned specificity in the third position. This must happen in such a way that the genetic code is instantaeously decodeable at each stage. In order that this be so, the Kraft Inequality, which is the necessary and sufficient condition that a code be instantaneous, must be satisfied at all stages of the evolution, both for the first extension and for the second extension. It is easy to see that the first extension code in Table 7.2 satisfies the Kraft Inequality:

$$1 \geq \sum_1^q r^{-l_i} = \sum_1^{16} 4^{-2} = \frac{16}{16} = 1. \tag{7.8}$$

In addition, the reader can easily verify that the modern second extension triplet code also satisfies the Kraft Inequality so that the genetic code is at all times instantaneously decodeable.

Instantaneous codes are well known to information theorists (Chaitin, 1975b; Hamming, 1986). Cullmann and Labouygues (1985) first proved that the genetic code is instantaneous. This is an important property, as, if the genetic code were not instantaneous, a decoding device would be needed to record the message before decoding could proceed. The code words of the Morse Code are not instantaneous so that the telegrapher must leave a letter space between each letter and a word space between words. Thus, the Morse Code is effectively a quaternary code.

7.4.2 The genetic code is optimal

The genetic code has the property of being *optimal*, which means that the genetic code employs the most economical use of its nucleotides. An optimal code is defined to be one that is both instantaneously decodable and that has the minimum average code word length. Khinchin (1957) proved that the maximum compression of a code is:

$$\frac{H(p)}{\log_2 q}, \tag{7.9}$$

where $H(\text{p})$ is the entropy of the probability vector p and q is the number of letters in the alphabet (see the Mathematical Appendix). In the case where the codons are all equally probable, it is easy to show that as the transition from a first extension to the third extension proceeds, the value of the Expression 7.9 gradually increases from two to three so the genetic code continues to be both instantaneous and optimal. This was first proved by Cullmann and Labouygues (1985).

7.4.3 Did the genetic code evolve in such a way that similar amino acids have similar codons?

We now have the means to determine quantitatively whether similar amino acids have similar codons. There are several pairs of codons that have a small BH distance and yet are at a Hamming distance of two and three. For example, the Pro-Gly BH distance is 0.01 but the Hamming distance is two and three; the Thr-Gly BH distance is 0.05, with Hamming distance two and three; the Asn-Asp BH distance is 0.05 with Hamming distance two and three. The Hamming distances and the BH distances for each pair of amino acids are given in Table 6.2 for comparison of those cases where mutationally related amino acids also have related codons and cases where this is not true.

However, all the amino acids that, in the second extension, were assigned triplet codons derived from the doublet codons of the first extension, according to the Markov evolution scenario described in Section 7.2, have codons that are at a Hamming distance of one. There also is considerable mutational similarity as shown by the BH distances. The BH distance between Cys and Trp is 0.07, the Hamming distance is 1. The BH distance between Asn and Lys is 0.09, the Hamming distance is 1. The BH distance between Gln and His is 0.23, the Hamming distance is 1,2. The BH distance between Asp and Glu is 0.02, the Hamming distance is 1. Leu captured UUA and UUG from Phe: The BH distance is 0.59 and the Hamming distance is 1,2. Arg and Ser divide the AGN codons and the BH distance is 0.26. Ile retains three of the AU codons in the second extension and Met is assigned AUG. The BH distance between Met and Ile is 1.00. Note that these BH distance figures are diameters and are rather small compared to the radii R in Table 6.3. Therefore, we see that the amino acids that lie in the column labeled *Second Extension* in Table 7.2 are only one Hamming distance apart from those found in the same row in the third column. The amino acids in the same row in the central and in the third columns have a small BH distance and do indeed have similar codons.

7.4.4 Are there really only twenty amino acids incorporated directly?

Much of the early considerations and speculations about the genetic code regarded the four nucleotides and the twenty amino acids as "magic numbers" (Crick, Griffith, and Orgel, 1957; Gamow, 1954a, 1954b). Subsequent research has shown they are not. Although twenty amino acids are known to participate in the first steps of protein synthesis, some 140 other amino acids are found in various proteins (Uy & Wold, 1977). These supernumerary amino acids are believed to be derived by a chemical pathway from one of the twenty amino acids incorporated directly in protein and are thus "posttranslational" or "derivatized." The information for the enzymes that catalyses these pathways is incorporated in the genetic message. Supernumerary amino acids could have been added at least up to 61 without going to the 256 codons of a third extension of the alphabet. Because DNA provides plenty of information capacity to accommodate the enzymes needed to change the amino acids directly incorporated to one of the 140 others found in modern proteins, this means of doing so may be often preferred to the more drastic path of revising the genetic code or going to a third extension. This is comparable to a computer programmer who adds a subroutine to a program to incorporate a new capability instead of completely rewriting the original program.

Zinoni et al. (1986), Chambers et al. (1986), Sunde and Evenson (1987), Zinoni et al. (1987), Mullenbach et al. (1987), Leinfelder et al. (1988), and Mizutani and Hitaka (1988) have found that an in-frame UGA non-sense stop codon can sometimes direct the incorporation in glutathione peroxidase of phosphoserine, which is then changed to selenocysteine. Berry, Banu, and Larsen (1991) have found that the mRNA for the type-I iodothyronine deiodinase, that converts thyroxine to 3,5,3′-triiodothyronine, contains an inframe UGA that translates selenocysteine. Pyrrolysine a derivative from lysine, is incorporated by the UAG codon usually a stop codon (Srinivasan et al. 2002; Hao et al. 2002). This shows that there are twenty-two proteinous amino acids and there may be more.

The point that twenty is not a magic number is further demonstrated by the fact that certain amino acid analogues are incorporated directly in the protein sequence. For example, Wong (1983) has reported a serial mutation in a tryptophan auxotroph of *Bacillus subtilis* strain QB928 to yield strain HR15 that grows well on four-fluorotryptophan but marginally on Trp. In the transition from QB928 to HR15, the replacement of Trp by four-fluorotryptophan changed the growth rate by a factor of 2×10^4 in favor of four-fluorotryptophan. Noren et al. (1989) have developed a site-specific method of incorporating unnatural amino acids into proteins by the use of a chemically acylated suppressor tRNA in response to a stop codon that had been substituted for the codon at the site of interest. This is essentially a method of assigning specificity to a codon previously unassigned. There is nothing in the theory presented in this book that is inconsistent with that and, indeed, the theory may be said to predict that more directly incorporated amino acids will be found, or perhaps could be created by genetic engineering.

7.5 The mitochondrial genetic codes, the endosymbiotic theory, and a common genetic code bottleneck

In this Section, I shall examine the characteristics of the mitochondrial codes in order to evaluate the endosymbiotic theory of mitochondria (Margulis, 1970). Mitochondria are organelles in the cell that have a protein synthesizing capability that is physically and genetically separate from the cytoplasmic system. Ten to twenty proteins are encoded in the genetic message in the mitochondrial DNA. According to the endosymbiotic theory, mitochondria were free-living bacteria at an early time in the history of life that were absorbed by eukaryotic organisms and now survive in an endosymbiotic relationship.

As the codes approached the bottleneck, they perforce became more nearly alike. This is reflected in the fact that there are only a few differences between the standard genetic code and the mitochondrial and other nonstandard codes. We find this is so in examining the codon assignments shown in Table 7.2 of mitochondria in *Mycoplasma caprolium, Tetrahymena, Paramecium,* yeast, and perhaps other organisms. The mitochondrial genetic codes differ from the standard genetic code and, indeed, from the codes of other mitochondria. These differences between the standard genetic code and the nonstandard code reflect the fact that they are not a property of a stereochemical relation between the codon and the amino acid as was once believed (Lacey and Mullin, 1983; Pelc and Welton, 1966; Welton and Pelc, 1966; Woese, 1967). This is demonstrated, for example, by the curious assignment of the CUN codons to Thr in yeast mtDNA rather than to Leu (Bonitz et al., 1980; Heckman et al., 1980; Li and Tzagoloff, 1979; Sibler et al., 1981). As another example, the use of UGA as a Trp codon in yeast mitochondria is reported by Macino et al. (1979) and in rabbit reticulocytes by Geller and Rich (1980). Furthermore, UGA codes for Trp in animal, fungal, and protozoan mtDNA, whereas UGG codes for Trp in plant mtDNA (Hiesel and Brennicke, 1983). The AUG, AUA, and AUU codons are initiator codons in HeLa mtDNA (Anderson et al., 1981; Montoya, Ojala, and Attardi, 1981) and in mouse mtDNA (Bibb et al., 1981).

Other deviations from the standard code are exhibited by different assignments of specificity to the third nucleotide in sense codons (Anderson et al., 1982). The readthrough or suppression of non-sense codons (Watson et al., 1987) is seen in this scenario as simply the assignment of specificity to non-sense codons in the same manner as specificity has been assigned to other codons. Although they have a release factor the non-sense codons are not special, *per se.*

In addition to the differences from the standard genetic code in mitochondria, Muto et al. (1985) and Yamao et al. (1985) have found that UGA codes for Trp in the small prokaryote *Mycoplasma capricolum.* Jukes (1985) suggested a mechanism for how this may have occurred. He outlined a series of evolutionary steps that could have lead to the replacement of the UGG codon by UGA.

The proposal discussed in this chapter shows that, assuming several different independent origin of life events, the first extension codes of such independent events would have become similar, but not necessarily identical, as they approached the bottleneck at code saturation. Perhaps there was a second origin of life event but the random walk (see Mathematical Appendix) took separate paths and several nearly similar genetic codes arrived at the bottleneck. Thus, several organisms may have emerged from the bottleneck at nearly the same time with genetic codes that differed only in a few assignments, especially in the third position. Notice in Table 7.2 that most of the differences in the mitochondrial codes are, indeed, in the third position (Anderson et al., 1982). This may mean that these differences were established upon emergence from the bottleneck at the extension from the doublet code to the triplet code and that the bacteria that were to become mitochondria had independent genetic codes.

Crick (1981), in one of those marvelous intuitions that have led him to so many discoveries, and without going through the mathematical argument above, has proposed: "What the code suggests is that life, at some stage, went through at least one bottleneck, a small interbreeding population from which all subsequent life has descended.... Nevertheless, one is mildly surprised that several versions of the code did not emerge, and the *fact that the mitochondrial codes are slightly different from the rest supports this*" (italics mine).

7.6 Summary of Chapter 7

The proposal for the evolution of the modern genetic codes discussed in this Chapter is based on the first principles of coding theory. It provides an explanation of the characteristics of the standard genetic code and other non-standard genetic code assignments. The discussion in Section 7.2 is based principally on the mathematics of coding theory and is not anecdotal or *ad hoc*. It enables us to see quite far back in the evolution of the genetic code and therefore of life. Lys with two codons is more frequent in modern proteins than Arg with six codons. Thus, six codons for Arg are selectively unfavored in the modern world. According to the comment in Section 7.3.1, one may suppose that these assignments were fixed prior to the incorporation of Lys in the protein alphabet. A change of assignment in the second extension is indistinguishable from genetic noise and is selectively prohibited.

The nature and the means of the assignments in the genetic code are best understood by comparison with other codes under the guidance of coding theory. For example, it is well to recall the discussion regarding the construction of block codes in Section 4.3.2. Given a binary alphabet of (0,1), we need to construct a code for a destination quaternary alphabet U, G, A, C. This can be done with the first extension, using all pairs of (0,1), namely, 00, 01, 10, 11. However, this code is saturated and provides no error protection. Therefore, the writer of the code may wish to go to the second extension and use all eight triplets of (0,1) and so on to further extensions as required.

In the case shown in Table 4.1, the code is written in the fifth extension and provides a full quaternary alphabet at the destination with considerable error protection. By the same token, it is possible to go to further extensions and provide for a larger alphabet at the destination and as much protection from error as desired. Presumably the evolution of the genetic code has followed the same coding theory procedure.

In the period when the genetic code was not saturated there were enough non-sense codons to facilitate evolution from unassigned to assigned codons. When nearly all codons had been assigned to amino acids and the codes were nearly saturated, changes in the codon assignments would be indistinguishable from noise. Therefore, the possibility of further evolution of the genetic codes was greatly diminished. This may explain why the standard, mitochondrial, and other genetic codes have remained unchanged since this early period in the evolution of life.

8

Haeckel's *Urschleim* and the role of the Central Dogma in the origin of life

I am in point of fact, a particularly haughty and exclusive person, of pre-Adamite ancestry descent. You will understand this when I tell you that I can trace my ancestry back to a protoplasmal primordial atomic globule. Consequently my family pride is something inconceivable. I can't help it. I was born sneering.

Pooh-Bah, *The Mikado*, Gilbert and Sullivan (First performed at the Savoy Theatre, London, England, March 14, 1885)

8.1 Haeckel, Pasteur, and speculations on the origin of life in the nineteenth century

The origin of life has concerned philosophers, poets, and theologians since antiquity. With due respect to the early experiments, when chemistry had not yet emerged from alchemy, of Francesco Redi (1626–98) and Lazzaro Spallanzani (1729–99) on spontaneous generation, Louis Pasteur (1822–95) showed that properly sterilized cultures remained so, for years, without germs added from without (Pasteur, 1848, 1922a, 1927b). Reductionists have not challenged this in spite of considerable effort.

8.1.1 Haeckel's protoplasmal primordial atomic globules

Ernst H.P.A. Haeckel (1834–1919) claimed priority for the nineteenth-century notion that life originated by self-organizing biochemical cycles from colloids or coacervates generated from organic substances, *Urschleim* (primeval slime) in the early ocean. He assumed that the early ocean served as a vat containing the *Urschleim* of the prebiotic Earth, where chemical evolution and its putative consequence, life, arose spontaneously *in flagrante delicto* from this nonliving matter and that it would almost inevitably arise on "sufficiently similar young planets elsewhere" (Bada and Lazcano, 2002a; 2003; Calvin, 1961; Miyakawa et al., 2002; Ponnamperuma, 1983; Rasmussen et al., 2004; Schopf, 1999; Simpson, 1964; Wills and Bada, 2000). (This speculation [the authors call it a theory] is known as the *prebiotic soup theory*. They are unaware that it is due to

Haeckel and they attribute it to Haldane [1929, 1954] or to Oparin [1938, 1957].)

Rarely mentioned is the paradox that aminoacyl-tRNA synthetases (aaRSs) are required to attach specific amino acids to their cognate tRNAs. These enzymes generate the reaction network, and the network is required to synthesize the enzymes and their component amino acids.

Haeckel maintained that Pasteur, in his sterilized culture experiments, had settled the negative only in certain circumstances. It being very difficult or impossible to prove a negative, many scientists in the nineteenth century, and many today, support spontaneous origin of life by chemical evolution in the early ocean. Haeckel (1905) wrote:

> The monistic hypothesis of abiogenesis, or autogeny in the strictly scientific sense of the word, was first formulated by me in 1866 in the second book of the *General Morphology*. (Haeckel, 1866) [Today autogeny is called self-organization.]

Haeckel was regarded as the German Darwin. His views were well known in the nineteenth and early twentieth century, for he was widely published in professional books and journals and in best-selling popular books translated from the German to several languages. As Haeckel (1905) explained:

> The chemical processes which first set in at this stage of development must have been catalysis, which led to the formation of albuminous combinations, and eventually of plasm. The earliest organisms to be thus formed can only have been plamodomous Monera, structureless organisms without organs; the first forms in which living matter individualized were probably homogeneous globs of plasm, like certain of the actual chromacea (*chrococcus*). The first cells were developed secondarily from these primitive Monera, by separation of the central caryoplasm (nucleus) and peripheral cytoplasm (cell body).

8.1.2 Haeckel and Pooh-Bah's genealogy

Haeckel's discussion of the origin of life from *protoplasmal primordial atomic globules* in the early ocean was so well known among scientists, theologians and the theater-going public in 1885 that Sir William Schwenck Gilbert (1836–1911) had Pooh-Bah, a comic, greedy, and conceited character who held all the offices (with the salaries) except Lord High Executioner, in the fictional Japanese town of Tititpu, introduce himself as being descended from a *protoplasmal primordial atomic globule*, no doubt from the *Urschleim*.

Of course, Pooh-Bah's genealogy would not have been funny, nor could Sir William have put these words in the mouth of Pooh-Bah, if his audiences in 1885 had not been familiar with "*protoplasmal primordial atomic globules*." (Part of the joke was that at the time the atomic nature of matter was a question of speculation.)

Sir William Gilbert's purpose behind Pooh-Bah's speech was to ridicule the hereditary in-bred aristocracy, whose pedigree was their only claim to their place in society, by one-upping them with a comic, disreputable character who could trace his ancestry *before* Adam.

How could the intellectually elite be unaware of Pooh-Bah's genealogy and its contribution to the origin of life? *The Mikado* is easily the most popular piece of musical theater ever written in the English language. There was a time when performances of this work were on the stage somewhere in the English-speaking world every day of the year. The late Thomas Hughes Jukes (1906–99) and I seem to be the only ones to have noticed Sir William Gilbert's contribution to the origin of life (Jukes, 1997; Yockey, 1992, 1995, 2000).

8.1.3 Racemic Urschleim, laevo amino acids, dextro ribose sugars, and the two-headed coin problem

Protein is composed of L-alpha amino acids, except glycine that is symmetric; sugars in DNA and RNA are dextro. But those compounds, if they were in the *Urschleim*, would have been racemic, composed of equal amounts of each handedness. Could abiotic chemical evolution select *only* the L-amino acids and *only* the dextro sugars leading to complex sequences and to life by abiological chemical evolution? Jeffrey L. Bada (1997) suggested that homochirality is simply a matter of chance and presumably he teaches that to his students.

The selection by chance may be illustrated according to the following fable. Suppose a practical person, a True Believer (Hoffer, 1951), and an independent person observe a coin tossed by the True Believer's Guru. The Guru's purpose is to demonstrate his triumph of Mind over Matter. Before the first toss, each trusts the Guru and believes that the coin is fair, although it hasn't been examined. If the coin is fair, the probability of heads or tails is one half and after ten tosses the number of heads and tails should be close to five each. After the appearance of the tenth head the practical person becomes suspicious that the coin is two-headed. The Guru continues to toss the coin one hundred times and finds heads each time. In less than two hours the Guru has selected one sequence from a total $2^{100} = 1.2676506002 \times 10^{30}$, a rather impressive feat! (The age of the universe is now believed to be 13.4 ± 1.6 billion years [Lineweaver, 1999]. There only about 4.225×10^{18} seconds since the origin of the universe.) Thus, the Guru demonstrates his mastery of Mind over Matter to the True Believer, or to the practical person that the coin is two-headed. (We shall find later in Chapter 11 that the sequence of all heads is not random, because it can be written by an algorithm shorter than the sequence.)

People who do not understand probability often say that extremely improbable events occur frequently, by a stroke of luck, in card playing, the sequence of automobiles on a highway, the seating of the audience at sports events and so forth (Dawkins, 1996). The question is, how much do we know *in advance* about the appearance of these events? We must calculate the probability of the event *before* it happens. It is wrong to say that *after the event* the probability is *one*, that is *certainty*.

Hidden among all the $1.2676506002 \times 10^{30}$ events is the beginning of the Code of Hammurabi written in the binary ASCII code. The beginnings of the lost plays of Sophocles are also among these events written in the binary ASCII code. But there is not enough time in the history of the universe for one to find them.

Thus, in order for Haeckel's paradigm to produce the protobiont in the *Urschleim*, it is first necessary for all the amino acids to have been of the laevo form, or perhaps, on another world, of all dextro-alpha amino acids. At the same time, the *Urschleim* must have had only dextro ribose sugars to produce RNA and DNA. Protein sequences must be of one handedness, are often much longer than one hundred amino acids, and so the prospect of their appearing in the early ocean, even with the help of a Guru, is beyond belief, at least for practical persons. As Louis Pasteur (Pasteur, 1848, 1922) showed in 1848, life is the only means capable of selecting molecules of only one-handedness.

Carl Sagan (1934–96) is well known as an advocate of intelligent life on Mars, Europa, and indeed in the universe. He wrote a book collaborating with the Russian author Iosef Samuilovich Shklovskii (1916–85) titled

Intelligent Life in the Universe, published in 1966. The book incorporates a translation, extension, and revision of Shklovskii's book *Universe, Life, and Mind*. I seem to be the only one who has noticed that, in Chapter 17, which appears to be from Shklovskii, the experiments by Pasteur with tartaric acid are discussed. Tartaric acid and indeed all biomolecules formed by natural means are racemic, composed of strictly equal amount of L and D forms. Shklovskii remarks that any racemic mixtures of organic compounds found on Mars could not be evidence of life. Furthermore, he notes that the organic molecules formed in the Miller–Urey experiments are racemic as were all organic molecules synthesized on the primitive Earth. The parable of the two-headed coin shows that no natural chemical procedure exits to form an optically active biochemistry.

The question of homochirality is ever present in the background of discussions on the origin of life (Bada, 1997). But as Mark Twain said about the weather, nobody does anything about it. Like a drowning man, people in the origin of life field grasp at any straw.

The genome is capable of directing the formation and incorporation of D-amino acids. The antibiotic gramicidin A is a linear pentadecapeptide that has an alternating D and L alpha amino acid sequence (Lang et al., 1991; Martinac and Hamill, 2002). It is isolated from *Bacillis brevis*. D amino acids occur in bacteria and also in mammals. High levels of D-serine are found in the mammalian brain where it is formed by the enzyme, serine racemase, that catalyses the formation of D-serine from L-serine (Stevens et al., 2003; Wolosker, Blackshaw, and Snyder, 1999); Physiological and genetic data in the human gene G72 and the gene for D-amino acid oxidase are implicated for schizophrenia Aerssens et al. (2002). This disease affects almost 1 percent of the world's population (Feng et al., 2002).

The dialectical materialist notion of Friedrich Engels (Chapter 9) that life is the existence of protein bodies and was bound to emerge from the *Urschleim* is one of the more distracting red herrings in the origin of life field (Bada and Lazcano, 2000a, 2002b, 2003; de Duve, 1991, 1995; Engels, 1954; Miyakawa et al., 2002; Schopf, 1999; Wills and Bada, 2000). In Section 6.4, I calculated the number of iso-1-cytochrome c sequences to be $6.42392495176 \times 10^{111}$. Thus, doing the problem correctly, using the Shannon-McMillan-Breiman Theorem, one finds that the 1-iso-cytochrome sequences are only a very tiny fraction $6.18601471259 \times 10^{-36}$ of the total possible sequences (Section 4.1). Consequently, it is quite out of the question that life emerged by chance from the *Urschleim*.

All speculation on the origin of life on Earth by chance can not survive the first criterion of life: proteins are left-handed, sugars in DNA and RNA are right-handed. *Omne vivum ex vivo.* Life must come only from life.

8.1.4 Darwin's views on the origin of life

Charles Robert Darwin (1809–82) is one of the saints of biology and of science in general. He lived at a time when most scholars and scientists were gentlemen. They did not work in teams and did not have to spend much of their time scrambling for grants from the politicians who administer the funding agencies.

Christopher Wills and Jeffrey Bada (2000), Christian de Duve (1995), J. William Schopf (1999), and a number of others quote Darwin's writings to support their position even though the quotation selected does not apply, is obscure or out of context. A favorite is from a private letter Darwin wrote to his friend Sir Joseph Hooker (1817–1911) in 1871. It appears in a footnote of Sir Francis Darwin, (1848–1925):

> It is often said that all the conditions for the first production of a living organism are now present, which could ever have been present. But if (and oh! what a big if!) we could conceive in some warm little pond, with all sorts of ammonia and phosphoric salts, light, heat, electricity, &c., present, that a proteine [*sic*] compound was chemically formed ready to undergo still more complex changes, at the present day such matter would be instantly absorbed, which would not have been the case before living creatures were found.

This passage (obviously not having been subjected to an editor's blue pencil) was not indexed and remained unnoticed until 1950 (see Hardin).

It is irresponsible and dishonest to reference this "warm little pond" quotation (Darwin, 1898) from Darwin's private correspondence as representing his views on the origin of life (Yockey, 1995, 2002). Everyone has the right to float tentative ideas and even nonsense to friends in his or her personal correspondence without responsibility being assumed by snoopers.

Darwin took this stance about the origin of life in the following quotation that appears on the *same page* as that quoted earlier.

> But I have long regretted that I truckled to public opinion, and used the Pentateuchal [The first five books of the Hebrew Scriptures,

classically thought to have been due to Moses] term of creation, by which I really meant "appeared" *by some wholly unknown process.* It is *mere rubbish*, thinking at present of the origin of life; one might as well think of the origin of matter. (Darwin, 1898, my emphasis)

Darwin did not believe in a "warm little pond" from which life is often alleged to have emerged by "chemical evolution" (Yockey, 1995, 2000, 2002a). Had he thought the "warm little pond" idea worthy of publication, he certainly would have done so. Significantly, Darwin avoided the origin of life controversy in Chapter XV of the sixth edition (1872) of *The Origin of Species*:

> It can hardly be supposed that a false theory would explain, in so satisfactory a manner as does the theory of natural selection, the several large facts above specified. It has been objected that this is an unsafe means of arguing; but it is a method used in judging the common events of life, and has often been used by the greatest natural philosophers. The undulatory theory of light has thus been arrived at.... [Darwin is showing the depth of his scholarship. He is referring to the work of Thomas Young (1773–1829). Young's famous double slit experiment proved that light is a transverse wave motion and that the waves going through the two slits interfere to form a pattern on a screen. This was contrary to Newton's particles of light. This problem was not resolved until the appearance of Max Planck's quantum theory that represents light as *hybrid* between a particle and a wave.] Darwin continues: "... and the belief in the revolution of the earth on its own axis was until lately supported by hardly any direct evidence." [Darwin again shows his wide knowledge of the science of his day. He is referring to the Foucault pendulum, invented by Jean Bernard Leon Foucault (1819–1869). In 1851, he suspended a 28 kilogram cannon ball by a 67 metre wire from the dome of the Pantheon in Paris. The floor of the Pantheon turns slowly under the pendulum, thus demonstrating the rotation of the Earth. The experiment provides an accurate measure of latitude. This demonstration settled for all time Galileo's support for Copernicus' theory that the Earth and all the planets, as well as their satellites, rotate against the field of fixed stars. (This was still an issue of biblical inerrancy in Darwin's time.)

Darwin draws the following conclusion about the origin of life:

> It is no valid objection that science as yet throws no light on the
> far higher problem of the *essence or origin of life.* Who can ex-
> plain what is the *essence of the attraction of gravity*? [An expla-
> nation of the *essence of the attraction of gravity* had to wait for
> Albert Einstein (Einstein, 1915a, 1915b, 1915c) and the expanding
> universe.] No one now objects to following out the results conse-
> quent on this unknown element of attraction; not withstanding that
> Leibnitz formerly accused Newton of introducing "occult qualities
> into philosophy." (Darwin again shows that he was well read in the
> scientific and philosophical literature of his day; my italics)

This paragraph shows that Darwin is one of the leading scholars in the
history of science. He believed that life *appeared by some wholly unknown
process*, and therefor is *undecidable.* I shall discuss the question what is
knowable and *decidable* in science and mathematics in Chapter 11.

8.1.5 Jacques Loeb's views on the origin of life

Long before the discovery of the genetic code, Jacques Loeb (1859–1924)
objected to the proposal that life emerged from colloids through catalysis.
Loeb (1924) was an expert in the chemistry of colloids and a very famous
man in the first half of the twentieth century. He was recommended for
the Nobel Prize numerous times but never received it. He is the model of
Dr. Max Gottlieb in Sinclair Lewis' novel *Arrowsmith.* In his book, *The
Dynamics of Living Matter*, published in 1906, Loeb wrote:

> But we see that plants and animals during their growth continually
> transform dead into living matter, and that the chemical processes
> in living matter do not differ in principle from those in dead matter.
> There is, therefor, no reason to predict that abiogenesis is impossi-
> ble and I believe that it can only help science if younger investiga-
> tors realize that experimental biogenesis is the goal of biology. On
> the other hand, our lectures show clearly that we can only consider
> the problem of abiogenesis solved when the artificially produced
> substance is capable of development, growth, and reproduction. *It
> is not sufficient for this purpose to make protein synthetically, or
> to produce in gelatin or other colloidal material, round granules
> that have an external resemblance to living cells.* (Italics mine.)

Jacques Loeb rejected Pooh-Bah's *protoplasmal primordial atomic glob- ule* model of the spontaneous origin of life from nonliving *Urschleim*, how- ever much it may have appealed to his mechanist and reductionist philoso- phy. He saw that colloids and coacervates lack the characteristic chemical processes, namely enzymes, by which organisms make sugars, fats, proteins, and other molecules essential to their metabolism. Moreover, they have no genome to control the formation of these critical compounds. It is a travesty that Jacques Loeb's comments are not now mentioned in the literature.

8.2 The formation of biogenic substances in the silent electrical discharge

Electricity was beginning to be used in a large scale in commerce, in- dustry and science at the turn of the nineteenth to the twentieth century. Experiments in electrical discharges were very prestigious. Papers in the field were published in German technical periodicals such as *Berichte der deutschen chemischen Gesellschaft (Reports of the German Chemical Soci- ety), Biochemische Zeitschrift (Journal of Biochemistry)*, and the *Zeitschrift für Elektro-Chemie (Journal for Electro-Chemistry)*, as well as in the British *Journal of the Chemical Society.*

8.2.1 Walther Löb, Oskar Baudisch, and E.C.C. Baly et al. find amino acids and other biological compounds in the silent electrical discharge and by UV

Plants take substances such as carbon dioxide, nitrogen, and water from the inorganic world and transform them to the living protein, cellulose, sugars, starch, and other biological compounds, all with strictly one chirality. Thus one has an act of the life processes under our very noses, quite different from inorganic chemistry. The chemical stability of carbon dioxide and nitrogen requires an energy source to be involved in the assimilation of these molecules.

The field of electrochemistry was well established (Collie 1901, 1905) in the early twentieth century. Walther Löb (1872–1916) pursued the fixation of nitrogen by silent electrical discharges *(stille elektrische Entladung)*, for biochemical reasons, as early as 1904, when he was a *Privatdozent* (a licensed university lecturer) at the University of Bonn. His apparatus, known as an "ozonizer", Figure 8.3, which shows Löb's apparatus, is well described. Löb reported (1909a) that he had frequently smelled the unpleasant and characteristic odor of butyric acid during investigations of the behavior of nitrogen in the presence of simple organic compounds under the influence of the silent electric discharge.

Löb (1913) succeeded in his search for glycine, but only in a reducing atmosphere, and thus he was the first man to produce an amino acid in the classic "possible prebiotic reducing [*sic*] atmosphere" of carbon monoxide, ammonia and water by means of an electrical discharge. Upon reading Löb's papers, I found that he knew exactly what he was doing and why he was doing it. The first sentence in his 1913 paper announced the purpose of the work that led to the formation of glycine in the silent electric discharge:

> "The question of natural nitrogen fixation is especially interesting in that it presents the source of the first organic nitrogen containing product for the formation of albumin bodies." [My translation.]
> [*In dem Problem der natürlichen Stickstoff-Assimilation interessiert besonders die Frage nach dem ersten organischen stickstoffhaltigen Produkt, das den Ausgangspunkt für die Bildung der Eiweißköper darstellt.*]
> Löb concludes his paper by saying:
> There is no doubt that according to previous results the amino acid found here is glycine. [My translation.]
> [*Es besteht aber bereits nach den bisher erhaltenen Ergebnissen kein Zweifel, daß die enstandene Aminosäure das Glykokoll ist.*]
> Here, succeeding for the first time, an amino acid has been produced artificially from the input products of the natural synthesis, which in any case, in the simplest phase, plays a role in the formation of natural protein as the final products of the natural synthesis from carbonic acid, ammonia and water without application of other materials, purely through supplying a special energy form that remains in close connection with the radiation.
> [*Es ist hierdurch zum ersten Mal gelungen, aus den Ausgangsprodukten der natürlichen Synthese, der Kohlensäure, dem Ammoniak und dem Wasser ohne Verwendung anderer Stoffe lediglich durch Zuführung einer geeigneten Energieform die mit der strahlenden in engem Zusammenhang steht, eine Aminosäure künstlich herzustellen, welche jedenfalls als eine der einfachsten Phasen im natürlichen Eiweißaufbau eine Rolle spielt.*] [My translation.]

Bada and Lazcano (2003) deny that Löb was interested in the origin of life. How could Löb say more clearly that he was working on a "prebiotic" experiment to synthesize "prebiotic elements of protein"? Clearly Löb thought his discovery of the formation of biologically important substances such as glycine, formic acid, acetic acid, formaldehyde, butyric acid, fatty acids, and other compounds, by means of electrical energy, was significant to biology (Löb, 1905, 1906, 1907, 1908a, 1908b, 1908c, 1909b).

Löb did not have the sophisticated modern techniques to find organic substances such as two-dimensional paper chromatography available to later investigators. Nevertheless, by 1915 Walther Löb and other scientists working on the formation of organic substances in silent electric discharges had established the formation of the following compounds: glycine (1913 and 1915), butyric acid (1909), formic acid (1904), acetic acid, formaldehyde, and other compounds of carbon. These are the same compounds, especially together with the tarry residue, found forty or more years later by Stanley L. Miller (born 1930) and others in spark discharges and in the Murchison meteorite.

8.2.2 The life of Walther Löb

Walther Löb had climbed the difficult ladder of promotion in German universities in the Imperial Germany of Kaiser Wilhelm II. He had attained the distinguished position of Director of the Biochemical Department at the Rudolf Virchow Hospital in Berlin. I was curious to know why there were no more papers from Löb after 1915 when he was still a young man. Purely on speculation I sent a letter to Herr Direktor, Rudolf Virchow Krankenhaus in Berlin. I asked for any material they might have on Professor Löb, his birth date, date of his death, and so on. I was delighted to receive, by return mail, an obituary and his biography from the Institut für Geschichte der Medizin Freie Universität Berlin. Walther Löb's last two papers were received in the offices of *Biochemische Zeitschrift* on December 8, 1914 and published in 1915. He died in Berlin at the age of forty-four after a short and unspecified illness on February 3, 1916, leaving his wife Agnes and four daughters Ilse (fifteen), Gertrud (thirteen), Dora (eleven), and Eva (seven years of age).

A eulogy was published in *Chemiker-Zeitung* on February 12, 1916, praising his research, contributions to science, and expressing sorrow at the untimely loss to his family and friends. Löb's goal of achieving some insight in how plants fix nitrogen and carbon dioxide to form protein was not reached by his methods nor by the same methods repeated by Stanley Miller and Harold Urey forty years later.

Oskar Baudisch (1913) showed that amino acids are generated by UV also only in a reducing atmosphere. Baly et al. (1922) reported the formation of α amino acids by UV. The amino acids formed by electric discharge or UV do not include the essential amino acids, methionine, and cysteine; both contain sulfur. Furthermore, many proteins, such as cytochrome c and hemoglobin contain iron or other metals that are critical to their activity.

Professor Walther Löb worked on this problem all his too-short professional career. Although it proved to be barking up the wrong tree, to explain the assimilation of carbon dioxide and nitrogen in plants; nevertheless, it was worth doing once but not twice. There has been considerable interest in Germany about the injustice to Walther Löb and the great overestimation of the value of Miller's work (Ahnhäuser, 2003). Löb's priority in the electro-chemistry of the silent electrical discharge and exploration of any function it may have had in "prebiotic chemistry" must be recognized (Mojzsis et al., 1999; Yockey, 1997, 2002b).

8.2.3 The contributions of Harold Urey and Stanley L. Miller

The first thing anyone must do when starting a research project is to read the literature. I show in what follows that, as Hamlet said to Horatio, "More honored in the breach than in the observance" (Act I, Scene IV).

Stanley L. Miller (born 1930), a student of Urey, is usually given credit (Bada, 1997; Bada and Lazcano, 2002, 2003; Schopf, 1999) for being the first to generate amino acids in a prebiotic atmosphere (Miller, 1953). He was only a second-year graduate student in 1953 and accordingly more sinned against than sinning. Urey (1893–1981) was a Nobel Laureate. Their relation was roughly that of a second lieutenant and a five-star general.

My experience with ruling personalities was as a graduate student at the University of California, Berkeley, 1939–43. Ernest Orlando Lawrence (1901–58) presided at laboratory meetings from a massive red leather chair, used by him alone. No one dared to sit there even when he was absent. He once felt it necessary to improve my character by some very firm remarks. Very few experiments were done at the Radiation Laboratory without his approval. His obsession with improving the cyclotron without regard for what it was to be used for, resulted in his missing the discovery of induced artificial radioactivity. Frédéric and Irène Joliot-Curie had made that discovery in 1934. As a graduate student I never heard Lawrence's embarrassment mentioned. The cyclotron provided much more intense radiation than the sources available to Frédéric and Irène Joliot-Curie so the flood gates were open for the discovery radioactive isotopes. Lawrence's headlong rush to build more high energy machines was essential to creating the field of high energy physics today (Heilbron and Seidel, 1989; Herken, 2002).

The field of electrochemistry of gasses was well known to Professor Harold Urey (1893–1981) and others at the University of Chicago in the 1950s. He had published at least three papers on the chemical effects of electrical discharges in gas (Urey and Smallwood, 1928; Urey, Dawson, and Rice, 1929; Urey and Lavin, 1929). Urey (1952) relied heavily on the dialectical materialism in the books by Oparin (1938, 1957) and Bernal (1951) for his guiding insight (See Chapter 9). He did not mention Haeckel or Haldane (1929) or Haldane's reference to Baly et al. (1922). He did cite Glocker and Lind (1939), which would have led him to the papers of Walther Löb. In his 1952 paper, he made the following suggestions:

> It seems to me that experimentation on the production of organic compounds from water and methane in the presence of ultra-violet light of approximately the spectral distribution estimated for sun-light would be most profitable. The investigation of possible effects of electrical discharges should also be tried since electrical storms in this reducing atmosphere can be postulated reasonably.

Thus, although he had contributed to the chemical effects of electrical discharges in gas, Urey did not take the trouble to search the literature, or to direct Stanley Miller to do so, and of course no one would have been so rash as to make that suggestion. A search of the literature would have found that Walther Löb, Oskar Baudisch, and Baly et al. (1922) had established that glycine is formed by the silent electrical discharge and by UV light only in reducing atmospheres.

Miller's contribution was to repeat the experiments of Walther Löb, Oskar Baudisch and Baly et al. (1922) with more sophisticated modern techniques, not available to previous authors, such as two-dimensional paper chromatography and elution from Dowex-50. Chemists in Löb's time used the ninhydrin reaction that is specific for alpha amino acids. A pure blue color is developed by heating with triketohydrindene hydrate. The apparatus used by Miller to examine amino acid formation in nonliving systems may be compared with that of Löb, Figure 8.2. Miller (1955) found, in addition to several amino acids, a number of other organic compounds, including an embarrassingly large amount of formic acid, glycolic acid, lactic acid, acetic acid, several butyric acids, and urea. Formic acid is the active ingredient

Table 8.1. *Yields from sparking a mixture of* CH_4, NH_3,
H_2O, *and* H_2. *The percentage yields are based on carbon
(59 mmoles [710mg] of carbon was added as* CH_4*)*

Compound	Yield (μ moles)	%
Glycine	630	2.1
Glycolic acid	560	1.9
Sarcosine	50	0.25
Alanine	340	1.7
Lactic acid	310	1.6
N-Methylalanine	10	0.07
α-Amino-n-butyric acid	50	0.34
α-Aminoisobutyric acid	1	0.007
α-Hydroxybutyric acid	50	0.34
β-Alanine	150	0.76
Succinic acid	40	0.27
Aspartic acid	4	0.024
Glutamic acid	6	0.051
Iminodiacetic acid	55	0.37
Iminoacetic-propionic acid	15	0.13
Formic acid	2,330	4.0
Acetic acid	150	0.51
Propionic acid	130	0.66
Urea	20	0.034
N-Methyl urea	15	0.051
Total		15.2

in the stings of bees, wasps and ants; acetic acid is what makes vinegar so sour. Miller admitted (1955) that the amino acids produced are among the simplest and do not represent the composition in protein. Furthermore, as Walther Löb and others had found, more than forty years previously, and long before Miller (born 1930) was born, the primary result of the spark experiments was a tarry mixture in the bottom of the flask, left unanalyzed. It has gone largely unnoticed that both Löb's and Miller's spark discharge experiments produce a racemic *Urschleim*. That would not have affected Löb's results because glycine, the amino acid he found, is symmetric, see Table 8.1.

Miller's paper had some difficulty in being published, as often happens to efforts from graduate students. Bada and Lazcano (2003) described the efforts of Urey to lay the foundations for Millers's research and to use his influence to get the paper published. Urey refused to be cited as an author because he believed that would diminish the credit due to Miller.

8.2.4 The contributions of John Burden Sanderson Haldane

John Burden Sanderson Haldane (1892–1964), in his often quoted paper, *The Origin of Life* (1929), mentioned the work of Baly et al. (1922):

> Now when ultra-violet light acts on a mixture of water, carbon diox-ide and ammonia, a vast variety of organic substances are made, including sugars and apparently some of the materials of which proteins are built up. This fact has been demonstrated in the labo-ratory of Baly in Liverpool and his colleagues. In the present world, such substances, if left about, decay – that is to say, they are de-stroyed by micro-organisms. But before the origin of life they must have accumulated in a hot dilute soup.

Haldane (1954), whose speculations are always entertaining, proposed four suggestions for the origin of life:
  (1) Life has no origin.
  (2) Life originated on our planet by a supernatural event.
  (3) Life originated from "ordinary" chemical reactions by a slow evo-lutionary process.
  (4) Life originated as the result of a very "improbable" event, which was almost certain to happen given sufficient time, and sufficient matter of suitable composition in a suitable state. [Note: improbable events are not almost certain to happen.]
As Haldane (1954) observed:

> The critical event which may best be called the origin of life was the enclosure of several different self-reproducing polymers within a semi-permeable membrane.

8.2.5 Drawing false consequences from the spark discharges

Many scientists were taken in by the Miller–Urey experiments and were lead to believe that those experiments *proved* that the atmosphere of the early Earth must have been reducing (Miller and Orgel, 1974):

> Arguments concerning the composition of the primitive atmo-sphere are particularly controversial. We believe that there must

have been a period when the earth's atmosphere was reducing, because the synthesis of compounds of biological interest takes place only under reducing conditions. Miller and Orgel, 1974,. ch. 4,. p. 33

Weber and Miller (1981) were so sure of their description of the creation of amino acids to form the *Urschleim* that they favored their students with the following wisdom:

> In our opinion, the basic reason that amino acids were used was their abundance in the primitive ocean. Most prebiotic experiments produce good yields of amino acids relative to other classes of organic compounds. The relevance of these experiments to the primitive earth is supported by the amino acid abundances found in the Murchison and other carbonaceous chondrites.

The reverent words "we believe" are clearly based on faith and a *quasi-religiosity will to believe*. They are appropriate in religious apologetics but not in scientific literature.

The evidence for the constitution and evolution of the Earth and its atmosphere must come independently from astronomy and geophysical chemistry. Scenarios for the origin of life must accommodate this. Good science dictates that one would search for evidence from several *independent* databases of a primeval reducing atmosphere (Abelson, 1966). Certainly one should not come to a scientific conclusion for ideological or religious reasons. Mojzsis, Kishnamurthy, and Arrhenius (1999) remarked:

> However, it is now held highly unlikely that the conditions used in these experiments [silent electrical discharge] could represent those in the Archaen atmosphere. Even so, scientific articles still occasionally appear that report experiments modeled on these conditions and explicitly or tacitly claim the presence of resulting products in reactive concentrations "on the primordial Earth" or in a "prebiotic soup" (see, for example, Deamer, 1997; Smith, 1998, 1999). The idea of such a "soup" containing all the desired organic molecules in concentrated form in the ocean has been a misleading concept against which objections were raised early. (Sillén, 1965)

8.2.6 The simulation of natural lightning

Prebiotic electric discharge experiments are still being carried out by the use of a Tesla coil that can be held in one's hand in the same manner as in 1913

(Löb, 1913; Miller, 1953; Stribling and Miller, 1987). Corona discharge in putative prebiotic atmospheres is often stated to simulate lightning. Such small equipment simulates corona discharge, not the enormous transient energy in lightning that is measured in nanoseconds, not days.

It is curious, in this time of megaexperiments, that no one has proposed to repeat these experiments with much larger amounts of *Urschleim* and much longer times. Large flash x-ray machines can simulate a lightning discharge in real time, including ionization, ultraviolet, heat and shock environment. Such machines can generate 20 megavolts and a current of 800 kiloamperes in a pulse width of 40 nanoseconds. Exposure either to x-rays or to the electron beam may be obtained. The temperature in the ambient gas may reach $10,000°K$ and is quenched in several nanoseconds. This is comparable to the environment in a lightning flash (Essene and Fisher, 1986). Whatever chemistry occurs in a lightning flash is simulated very closely, combining simultaneously all the elements using a large x-ray machine mentioned above (Yockey, 1992).

8.2.7 Chemical evolution by the polymerization of amino acids on mineral surfaces

There is a considerable literature on the polymerization of nucleic bases on mineral surfaces, such as zeolites and feldspars, with the hope to provide a library of sequences and to bypass the difficulty of forming peptide bonds in an aqueous medium (Bernal, 1951, 1967; Cairns-Smith, 1965, 1971, 1982; Ferris et al., 1996; Popa, 1997; Smith et al., 1998; Smith et al., 1999; Sowerby and Heckl, 1998; Sowerby, Cohn, Heckl, and Holm, 2001). These authors suggest that the mineral surfaces provide a "library" of sequences that may lead to a "genetic takeover." Sowerby and Peterson (2002) present an outline of the objections for a *de novo* appearance of linear self-replicating polymers. They propose that free purine and pyrimidine bases, alone adsorbed on a suitable surface, could have constituted a primitive coding template for the construction of catalytic proteins. Even the best attempts report monomers fifty-five long. This is far too short to form a genome or even a protein. As Orgel remarked:

> One conclusion is that theories that involve the organization of complex, small-molecule cycles such as the reductive citric acid cycle on mineral surfaces make unreasonable assumptions about the catalytic properties of minerals. (Orgel 2000)

These proposals rely on Miller (1953) rather than Haeckel (1866) for their sources of amino acids (Sowerby et al., 2001; Sowerby, Peterson, and Holm, 2002; Sowerby and Peterson, (2002). Hazen et al. (2001) have proposed that the selective absorption of linear arrays of D and L amino acids on calcite represents a "plausible" geometric mechanism, with subsequent polarization, for the production of homochiral polypeptides on the prebiotic Earth. The values reported are very close to one and thus serve to contradict their conclusion.

8.2.8 Wächtershäuser's conjecture on the origin of life

Wächtershäuser (1994, 1997, 1988a, 1988b, 1988c, 1998a, 1998, 1990, 1997, 2000, 2003) is one of the important workers in the origin of life field (Hagmann, 2002). He is guided by the philosophy of Sir Karl Popper. Wächtershäuser's objections to Haeckel's *Urschleim* are much the same as mine.

Wächtershäuser has made a very ingenius proposal for the fixation of carbon and the generation of membranes of lipids and anionic peptides. Central to his speculation is the idea that early life was autotrophic (in contrast to the heterotrophy of the *Urschleim* paradigm and the adsorption of amino acids on clay). He proposed that the fixation of carbon proceeds directly on the surfaces of minerals such as FeS and produces iron pyrite FeS_2. This is the first source of energy for life. This process is described (Wächtershäuser, 1990) by the following exergonic reactions where $\Delta G°$ is the free energy for pH zero and $\Delta G°'$ is the free energy for pH 7:

$$4CO_2 + 7H_2 \rightarrow (CH_2\text{---}COOH)_2 + 4H_2O$$
$$\Delta G° = -151 \text{ kJ/mol}$$
$$4HCO_3^- + 2H^+ + 7H_2 \rightarrow (CH_2\text{---}COO^-)_2 + 8H_2O$$
$$\Delta G°' = -160 \text{ kJ/mol.}$$

The free energy is much larger in the presence of H_2S and FeS:

$$4CO_2 + 7H_2S + 7FeS \rightarrow (CH_2\text{---}COOH)_2 + 4H_2O + 7FeS_2$$
$$\Delta G° = -420 \text{ kJ/mol,}$$

and

$$4HCO_3^- + 2H^+ + 7H_2S + 7FeS \rightarrow (CH_2\text{---}COO^-)_2$$
$$+ 7FeS_2 + 8H_2O \quad \Delta G°' = -429 \text{ kJ/mol.}$$

These reactions provide the energy for the fixation of carbon and also a binding surface for the organic constituents that are formed. Any process by which the constituents lose their surface bonding is irreversible and the material is lost by dissolution in the ocean. The propensity for thermal degradation in a surface-bonded state is less than in solution so that the process may proceed at the higher temperatures typical of submarine hot springs (see Section 8.3.1).

Several forms of iron sulfides have been found in extremely primitive sulfur bacteria that live in anaerobic conditions (Farina et al., 1990; Mann et al., 1990; Williams, 1990). This seems to be a corroboration of Wächtershäuser's proposal (Popper, 1990; Russell, Hall, and Gize, 1990).

Among the advantages of his proposal are that it generates the organic molecules without concomitant large amount of tarry material that should have been absorbed in sedimentary rocks and become kerogen. It generates the compounds in place, thus avoiding the need for a method of concentration. It is consistent with energy requirements and thermodynamics. The *Urschleim* paradigm has a serious defect that requires that amino acids polymerize in aqueous solution, whereas peptides actually hydrolyze.

Rather than requiring ready-made "building blocks," Wächtershäuser's proposal assumes only the CO_2, H_2S, and FeS found in submarine hot springs. The reaction products remain in place and there is no requirement for concentration, a serious difficulty with the *Urschleim* paradigm. The reactions shown above are far from thermodynamical equilibrium and are irreversible as must be the case in living systems. Although he did not mention the papers by Schidlowski (1983a, 1988, 2002), those papers are among the strongest support for his speculation.

The proposal is incomplete in that it generates a two-dimensional monomolecular layer when all informational molecules are sequenced. It shares with other speculations the idea that life is just complicated mechanistic chemistry. I have not found any reference to the handedness of amino acids and the ribose sugars in DNA and mRNA. It appears to be "proteins first" (Huber and Wächtershäuser, 1997; 1998; Huber et al., 2003) and of course that is prohibited by the Central Dogma (Yockey, 1978, 1992, 2002 and Chapter 3).

He does not address adequately the generation of complexity and of genetic code (Orgel, 2000). Wächtershäuser (1994) believes that the use of the information metaphor has led to the unfortunate prejudice that life must have started with a polymer sequence:

By my theory, life would then have started with analog information, and it would have "invented" digital information later. (Wächtershäuser, 1994).
This view is not in accordance with my remarks in *Section 1.1.2.*
Wächtershäuser (1997) suggests that:
The theory of biological evolution is an historic theory. If we could ever trace this historic process backwards far enough in time, we would wind up with an origin of life in purely chemical processes.

That is well and good but the test of the pudding is in the eating. I have discussed (Yockey, 2000, 2002) in Chapter 11 *"The Unknowable"* that tracing this historic process backward in time to the origin of life is *unknowable* or *undecidable*.

Speculations based on the formation of sequences of molecules directly on the surfaces of minerals face the problem of the Guru and the two-headed coin. An unbelievably enormous number of sequences must be formed in order to find the ones that form the minimal essential genome of the first living organism essential genome of the first living organism (Section 6.4).

8.3 The origin of matter and the origin of life

The origin of matter occurred in the Big Bang 13.4 $\pm$ 1.6 billion years ago (Cayrel et al. 2001; Freedman and Feng, 1999; Krauss and Chaboyer, 2003; Lineweaver, 1999). Astronomers and physicists can trace the origin of matter to within the Planck time, about 10^{-34} seconds, after the *First Cause* that originated the fire ball of the Big Bang. Darwin's prescience in the origin of matter and the origin of life is extraordinary. Astronomers and physicists know more about the "origin of matter" than biologists know about the origin of life. The origin of life appears to be a more difficult problem!

8.3.1 The faint young Sun and Sun as a red giant

When life appeared on Earth the "faint young Sun" radiated only about 70 percent of the energy it now generates. The core of the Sun, at its present position in the Hertzsprung–Russell diagram (Figure 8.1), is gradually being enriched in helium. It is contracting and the temperature of the core is increasing where most of the thermonuclear reactions occur. The rate of thermonuclear reactions is very sensitive to temperature and this will cause

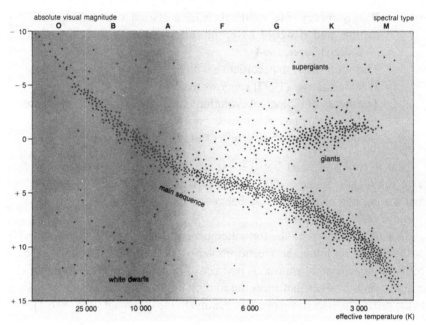

Figure 8.1. The Hertzsprung–Russell diagram. The absolute visual magnitude or luminosity of stars of known distance is shown plotted on the vertical axis. The effective surface temperature is plotted on the lower horizontal axis. The spectral type is given on the upper horizontal axis. The sun began its life in the gravitational collapse of a cloud of gas and dust. When thermonuclear ignition occurred the Sun was located at the lower right of this figure. As it became hotter and brighter it progressed to the left and up along the main sequence. It is now of the G2 spectral type. The effective surface temperature is about 6000° K. In about 10^9 years the energy received from the Sun will heat the Earth so that the oceans will boil. Eventually the Sun will become a red giant and its surface will extend to the present orbit of the Earth. Reprinted from Audouze, J. & Israel, G. (1985). *Cambridge Atlas of Astronomy*, Cambridge, New York: Cambridge University Press.

the insolation to increase about 1 percent per hundred million years (Gough, 1981). The Sun, progressing inevitably along its evolutionary path, before five billion more years will expand beyond the present orbit of the Earth and become a red giant (Penrose, 1989) like Aldebaran in the constellation Taurus (the Bull) or Betelgeuse in the constellation Orion (The Hunter).

The habitable zone is the space about the about the Sun within which the climate stability of a planet supports liquid water. As the Sun ages, and the insolation increases producing an increasingly effective greenhouse effect, the oceans will boil away and the Earth will be cooked into lifelessness like Venus today (Correia and Laskar, 2001; Kasting, 1993; Sagan and Mullen, 1972); Life has already existed on Earth for three quarters of its allotted time. Perhaps the last organisms will be the thermophilic bacteria that started the whole thing! *Sic transit gloria mundi!*

8.3.2 Warm little ponds, tidal lagoons on Earth, and the formation of the Moon

It is helpful for the reader to know the elements of current ideas in cosmology in order to understand the modern view of the conditions on the early Earth when life originated. The age of the Earth can be found from the age of the Allende CV chrondite. The radioactive decay periods of uranium to the lead isotopes $^{227}Pb/^{206}Pb$ serve as a clock which yields an age of 4.559 $\pm$ 0.004 billion years (Tilton, 1988). More recent values of the age of the earth, 4.5627 $\pm$ 0.0006 billion years, are in agreement (Amelin et al., 2002; Bizzarro et al., 2003).

Sidney Fox (1994, 1996) and others have proposed that life originated in "warm little ponds" or "the evaporation of tidal lagoons." The question of "warm little ponds" or "the evaporation of tidal lagoons" on the early Earth has been answered by the most widely regarded scenario for the formation of the Moon, namely, that it was formed within ten to thirty million years after the formation of the solar system (Kleine, Münker Mezger, and Palme 2002; Yin et al., 2002). A rare event, indeed!

The Moon is believed to have been formed by the off-center collision with the early Earth of a proto planet, Theia (In Greek mythology, $\Theta\varepsilon\iota\alpha$, the goddess of sight), a Mars-sized object (Halliday, 2004) (Cameron and Benz, 1991; see Cameron and Ward, 1976; Canup and Asphaug 2001; Canup and Righter 2000; Mojzsis et al., 1999; Shiguru et al., 1997; Wiechert et al., 2001). Münker et al. (2003) give a date for the final core-mantle equilibration on Earth of 4.533 bilion years ago.

Part of the debris from the impactor of this empyrean collision accreted to form the Moon in an orbit just beyond the Roche limit. The Roche limit is the distance to the center of a fluid body that has no internal strength to prevent disruption by the stress across that body caused by the planet's gravitational field. The Roche limit was about 2.86 times the radius of the Earth at the time, or about 71,500 miles–the Moon is now about 250,000 miles from the Earth.

This collision of a Mars-sized object on the early Earth shows that a *just-right* Goldilocks event may have played a decisive role in determining the "suitability" of the Earth for the origin of life. For example, the fact that Venus has no large satellite comparable to the Moon suggests that the

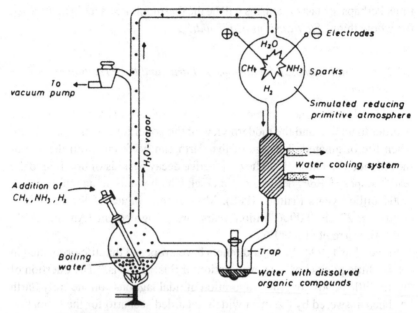

Figure 8.2. The Urey-Miller experiment. Water vapor and the added gases (CH_4, NH_3, and H_2) were cycled through an electric spark and cooled in a jacket. The condensates were collected in a trap, and the rest was recycled. Within a week at least four protein-forming amino acids had been manufactured and collected in the trap. (Schwemmier, 1984, p. 20, Figure 1.)

fickle goddess Fortuna contributed the essential event that distinguished the history of Venus from that of the Earth. Without this singular off-center collision event, by an object of just the right size, at just the right time, the Earth may have followed Venus to become a furnace rotating slowly backward (Correia and Laskar, 2001). This Goldilocks event, in which everything must be *just-right*, indicates that truly Earth-like planets may be very rare in the Milky Way Galaxy.

The lunar tides, tidal currents, and tsunamis, washing over the low continents, were enormously greater (up to 1–2 km) than those of today because the Moon was close to the rapidly spinning Earth and the day was approximately five hours (Canup and Asphaug, 2001; Canup and Righter, 2000; Kerr, 1999; Melosh, 2001; Popa, 1997). Consequently, there were no "tidal lagoons" or "warm little ponds" on the early continents where life could originate spontaneously by "chemical evolution" from a "primeval soup" as presumed by many speculations on the origin of life (Bada and Lazcano, 2002; Levy and Miller, 1998; Schopf, 1999). It is typical of speculations on

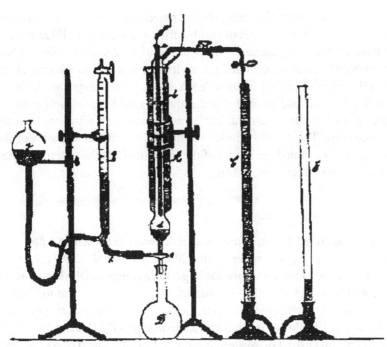

Figure 8.3. Apparatus used by Walther Löb for studying the biometric synthesis of organic compounds, including aldehydes (1906) and amino acids (1914) in cold plasmas, with CO_2, and NH_3 as reactants in various combinations. (A) plasma generator ["elektisato"]; a, inner tube; b, outer tube; c, plasma discharge space; d, liquid bulb; e, end three-way stopcocks; (B) buretter with levelling funnel; (D) recipient flask for liquid with reaction products; (C) and (E) gas burettes. (Reprinted from Löb, 1906).

the origin of life to be based on conditions on the early Earth that did not exist.

A cataclysm occurred between 4.0 and 3.8 Gyr years ago, when the lunar basins with known dates were formed. This led to the metamorphism of the lunar surface by the collision of an enormous number of asteroids and/or comets (Cohen, Swindle and Kring, 2000; Culler et al., 2000; Kerr, 1999; Nisbet and Sleep, 2001; Schoenberg et al., 2002). These collisions occurred in a brief pulse of about two hundred million years. The cataclysm produced more than seventeen hundred craters 20 to 1,200 km in diameter on the Moon.

It is widely believed (Schoenberg et al., 2002) that the Earth had a meteorite bombardment at the same time. Because the Earth has a larger cross-section and a larger attraction of gravity there may have been more than seventeen thousand large impact events in this time.

Knowledge about the impacts of bolide objects that created large craters in the Barberton greenstone belt of South Africa and the eastern Pilbara block of Western Australia comes from evidence of four large bolide impacts that created large craters, altered the environment, and deformed the target rocks (Byerly et al., 2002). The single zircon uranium lead ages are 3.470 ± 2 million years and are associated with sedimentation by an impact generated tsunami. In Western Australia, this is associated with a major erosional unconformity. The Apollo missions to the Moon returned samples of the lunar soil that enabled the dating of the lunar maria to thirty-five hundred million years (Culler et al., 2000).

8.3.3 The age of life on Earth

The Earth accreted from the first solids in the early solar nebula 4.567.2 ± 0.6 million years ago (Amelin et al., 2002). Caro et al. (2003) used the $^{147}Sm^{143}Nd$ chronometer to time the large-scale features of the Earth's mantle to 4.460 ± 115 Myr. The radioactive decay series of ^{176}Lu to ^{176}Hf is widely used to measure the age of the Earth's early mantle system (Bizzarro et al., 2003). The oldest crustal rocks surviving from the Hadean time are in the Murchison District of Western Australia, from the island of Akilia outer Gothåbsfjord in southwest Greenland and from Acasta Gneiss of northwestern Canada. These rocks contain zircons that can be dated by the radioactive decay of the 238 isotope of uranium to several lead isotopes, the known radioactive decay periods serving as a clock Amelin et al. (2002). Mojzsis et al. (2001) find these 4.300 Myr old zircons, from the Murchison District of Western Australia, formed from magmas containing a significant component of reworked continental crust that formed in the presence of water. This dates the earliest oceans at 4.300 Myr. Wilde et al. (2001), working with zircons from the Yilgarn Craton in Western Australia, found zircon grains 4.400 ± 8 Myr representing evidence for a continental crust and oceans on the Earth at this time.

Taking these dates as being well determined, one looks for specific evidence of life in these objects. Carbon compounds are composed of a light isotope ^{12}C and the much rarer ^{13}C. It has been known for some years that carbon compounds that are biologically processed have a significant fractionation for ^{12}C. This has served as an indicator of biological origin (Schidlowski, 2002). Fedo and Whitehouse (2002) dispute the identification of the quartz-pyroxene rock as banded iron formation (BIF). Their analysis identifies these rocks as ultramafic igneous, not BIF (Kerr, 2002). The biogenic origin of the graphite carbonate-rich in Isua was inferred from the assumption that these rocks had a sedimentary origin. However, recent field and laboratory investigations have shown that most if not all carbonate in Isua is metasomatic rather than sedimentary in origin (Van Zuilen. Lepland, and Arrhenius, 2002). Because water is widely believed to be necessary for life, the older papers can no longer be presented as evidence of an *Urschleim* in the early oceans on Earth or on "sufficiently similar young planets elsewhere." *The absence of evidence is evidence of absence.*

Schopf et al. (2002) have interpreted certain objects found in Precambrian cherts as being true microfossils. Other scientists (Brasier et al., 2002) disputed this identification in the same issue of *Nature*. Schopf depends on morphology in all of his analyses. Dalton (2002) reported that two of Schopf's collaborators have disputed Schopf's conclusions. Garcia-Ruis et al. (2003) have synthesized inorganic micron-sized filaments that show noncristallographic, curved helical morphologies, similar to the forms from the Precambrian Warrawoona chert formation in Western Australia, that Schopf et al. (2002) have proposed to be the oldest microfossils (Kerr, 2003). This interpretation may be compared to Rorschach inkblots. One is reminded of the case of the Allan Hills 84001 meteorite that I shall discuss in Section 8.3.4.

The evidence cited earlier is very far from the official NASA policy for the origin of life based on the Marxist dialectical materialism of Engels and Oparin discussed in Chapter 9. The determination of the time of the earliest appearance of life on Earth is still in a state of flux. For the time being, let us take 3.850 billion years as a working date for the appearance of life (Mojzsis and Harrison, 2002). This is about the time when the late heavy bombardment ceased (Kerr, 1999; Nisbet and Sleep, 2001).

8.3.4 Life on Mars?

President Clinton gave a press conference on the South Lawn of the White House on August, 7, 1996. He spoke about the meteorite Allan Hills 84001, or for short ALH84001. The president noted that ALH84001 had been picked up from a glacier in Antarctica in 1984. He discussed briefly the history of ALH84001 from its formation four billion years ago on Mars, its escape because of the shock from the impact of a body striking Mars, the sixteen-million-year journey through space and its arrival thirteen-thousand years ago on a glacier in Antarctica.

Mr. Daniel Goldin, NASA Administrator at the time, held a briefing on the "Discovery of Possible Early Martian Life" at NASA Headquarters in

Washington, DC. Mr. Goldin is well known for his enthusiasm for the search extraterrestrial intelligence, life on other planets, manned missions to Mars, and the search for "evidence" of life on Mars and Europa (Chyba and Philips, 2002; Kerr, 2001). President Clinton and Mr. Goldin have since gone on to other tasks in their careers.

As expected by the NASA public relations office, ALH84001 attracted the attention of *Newsweek* (1997). Considerable controversy was reported in support of the proposed "evidence" of life on Mars. The long story of life on Mars from ALH84001 (McKay et al., 1996; Thomas-Keprta et al., 2001) ended with the determination that the artifacts upon which this claim is made are terrestrial contamination (Bada et al., 1998; Bradley et al., 1997; Buseck et al., 2001; Jull et al., 1998; Kerr, 2002; Yockey, 1997d, 1998).

There are a number of traces of what may be water worn networks on a warm wet early Mars (Malin and Edgett, 2003). Valley networks on Mars cut across the heavily cratered southern highlands. These features were formed during the period of heavy bombardment in the early solar system that affected Earth and the Moon as well. Speculation about a warm wet early Mars is based on the belief that there was a long-lasting greenhouse climate and implies that Mars was teeming with life in that era.

On the contrary, the analysis by Segura et al. (2002) shows a cold dry planet, an almost endless winter broken by periods of scalding rains followed by flash floods. Carbonate minerals may provide evidence for the presence of liquid water on Mars in the remote past. They form readily in a carbon dioxide atmosphere in the presence of water. On Earth, water and atmospheric carbon dioxide form carbonic acid that combines with rocks and precipitates in lakes and oceans to form carbonate deposits like the White Cliffs of Dover. Observations by means of the Thermal Emission Spectrometer have provided measurements of the thermal emissivity spectra of the surface dust of Mars. Only trace evidence was found of carbonates, like limestone, in the mile-scale surface deposits. Only 2 percent carbonate dust was found, whereas wet weathering would produce 20 percent (Astrobiology.com, 2003; Bandfield et al., 2003; Kerr, 2003). The Thermal Emission Spectrometer has been used to search for the mineral olivine which is unstable in aqueous environment (Christensen et al., 2003). Hoefen et al. (2003) have found olivine in the Nili Fossae region of Mars. They believe that supports the argument that Mars has been cold and dry for a long time. No oceans on Mars means no *Urschleim* and no life. Traces of water ice on the poles of Mars do not make the cut.

If one believes that ALH84001 came from Mars, one must believe that meteorites are exchanged between Earth and Mars. The only uncontroversial evidence of life on Mars that did not come from Earth would be to find biological material that is *both* dextro in amino acids and laevo in nucleic acids. Unfortunately, amino acids and nucleic acids become racemic over the terrestrial geologic time scale of 5–10 million years. *The absence of evidence is evidence of absence.* That does not deter faith-based search for life on Mars by all space-faring nations.

The exploration of Mars has considerable international interest and political support, especialy from the life-on-Mars congregation (Morton, 2003). The British built *Beagle 2*, named after the ship on which Charles Darwin made his historic voyage, was programmed to touch down on December 24, Christmas Eve, 2003, at 9:54 P.M. Eastern Standard Time. Its purpose is to search for chemical signs of life in the Martian soil. No signal has been received from *Beagle 2* as this book goes to press (Butler, 2004). The European Space Agency *Mars Express* has gone into orbit at the same time for a two-year mission to map the planet surface. NASA's *Spirit* has landed on Mars and is sending pictures, as this book goes to press. Its twin, *Opportunity*, landed on the opposite side of Mars in January 24, 2004 and is sending pictures. They will mount geological surveys of sites that may bear traces of water. Japan has a spacecraft *Nozomi*, launched in July 1998. As this book goes to press, *Nozomi* has suffered numerous setbacks: fuel shortage and damage from a solar flare. That is the fourth failure in Mars missions in the last five years and reflects on the judgment of those who are promoting a manned mission. Those on a manned mission wish to come back, although I have a short list of some who could remain there and contribute to the cause.

Now that Venus has proved to be a furnace, Mars is of interest because it resembles Earth more than any other planet. The axis of the spinning Earth has varies from 22° to 25.5° from the plane of the ecliptic over the last ten million years. The range for Mars is 14° to 48°. The stabilization of the Earth is by the large Moon. Stabilization can not be provided by the two very small moons of Mars, Phobos, and Deimos (Head et al., 2003). Thus, Mars is a test object for study of the history of the Earth. Mars has had ice ages similar to those on Earth. Regardless of whether life originated on Mars, this effort is justified because the processes of planetary formation and the geological history are of important for comparison with that of Earth.

8.3.5 Exobiology and life from outer space

Directed Panspermia, the suggestion that life was introduced from outer space, raises its head from time to time. Hermann Ludwig Ferdinand von Helmholtz (1821–94) and Lord Kelvin, William Thompson (1824–1907), proposed the Directed Panspermia theory in the nineteenth century. Helmholtz and Svante Arrhenius (1859–1927) proposed that life is eternal and that meteors that roam about the solar system might contain germs of organisms that, under favorable conditions, might reach the Earth and other planets (Arrhenius, 1908).

Research on comets and carbonaceous meteorites, supported by the Exobiology Program (http://astrobiology.arc.nasa.gov) of the National Aeronautic and Space Administration (NASA); requires a genuflection to the origin and evolution of life (Cooper et al., 2001; Morrison, 2001). A collection of the earlier papers can be found in (Thomas, Chyba, and McKay, 1997). There is now a journal, *Astrobiology*, in which Morrison (2001) comes very close to violating the proscription on the establishment of religion in the First Amendment of the United States Constitution:

> As our understanding of living systems and the physical universe increases, we will confront the implications of this knowledge in more than just the scientific and technical realms. To understand the consequence will require multidisciplinary considerations of areas such as economics, environment, health, theology, ethics, quality of life, the sociopolitical realm, and education. Together we will explore the ethical; and philosophical questions related to the existence of life elsewhere, the potential for cross-contamination between eco-systems in different worlds, and the implication of future long-term planetary habitation and engineering. (Morrison 2001)

NASA is not responsible for: "... economics, environment, health, theology, ethics, quality of life, the sociopolitical realm, and education." Readers who wish to know "What is astrobiology?" will find the NASA Astrobiology Road map at http://astrobiology.arc.nasa.gov.

Finding the "seeds of life" in interstellar dust grains and ice analogs has been proposed as having been available in all planetary systems for the origin of life (Bernstein et al., 2002; Caro et al., 2002).

Amino acids are condensed into polymer chains by the peptide bond. The peptide bond is formed by a condensation reaction between the amino group of one amino acid and the carboxyl group of another. A molecule of water is expelled when the peptide bond is formed. Thermodynamics requires that this reaction proceed in the direction of hydrolyzing in an aqueous medium. This direction is opposite to that required to form polymers. Consequently, amino acids *will not* form sequences in the primeval ocean. I pointed out in Section 6.1.1 that the activity of proteins depends not only on the sequence of amino acids but, most important, on the active pocket that contains contains metal ions such as iron, zinc, copper, and magnesium (Thompson and Orvig, 2003). Thyroxine contains iodine. Furthermore, the sequence of amino acids must fold up properly to form a protein (Selkoe, 2003). This doesn't seem to be discussed in origin of life speculations (Rasmussen, 2004).

George Gaylord Simpson's (1902–84) remark on *exobiology* (Simpson, 1964) is pertinent today:

> There is even a recognition of a new science of extraterrestrial life, sometimes called *exobiology* – a curious development in view of the fact that this "science" has yet to demonstrate that its subject matter exists.

If life is a Cosmic Imperative (de Duve, 1991, 1995) and would appear everywhere conditions of physics and chemistry permit, such life may be quite different from that on Earth. Perhaps it would be based on dextro amino acids and laevo mRNA. Indeed, rolling back the history of Earth, we find that life has indeed been quite different in the past and has existed with a nonprevalence of humanoids (Simpson, 1964). There were five major mass extinctions, any one of which, had it been more severe, could have exterminated life altogether on Earth. One must presume that such events are not unusual on planets around Sun-like stars.

8.3.6 Concentration of amino acids and the origin of life from sea foam

Bernal (1951, 1967) and Banin and Navrot (1975) have suggested that one means of concentration of organic matter from an *Urschleim* is in the form of sea foam. Perhaps short chains of the compounds, generated in the *Urschleim*, floated to the surface and were driven by wind and waves to accumulate on the shores of the ancient seas. By this means, dilute amino acids and other biomolecules could have become very concentrated. According to these authors, there should have been no great difficulty in polymerizations and presumably the generation of a protobiont.

The credibility of this scenario is supported by the authority of Hesiod (Theogony c. 700 B.C.) who tells us that Aphrodite (her name means foam-born), arose from the sea foam on the island of Cythera. From there she was wafted on Homer's "wine dark sea" to Cyprus. However, it must be noted that her emergence was aided by the god Kronus who castrated his father Uranus (the god, not the planet) and threw the genitals into the sea near Cythera where they served as an *élan vital for* the emergence of Aphrodite. The goddess, being immortal, was able to escape the effect of the intense ultraviolet light on the Earth at the time and that would have exposed mortal organisms to a lethal dose in less than 0.3 seconds (Sagan, 1973). Aphrodite (Homer [850 B.C.], Euripides [c. 440 B.C.] is still widely worshiped today, especially near colleges, universities, and army camps. Nevertheless, as far as the emergence of mortal beings is concerned, one must be a bit more skeptical about this mode of origination, because we assume only the established processes of physics and chemistry and no *élan vital*. Let us, therefore, consider other means of concentration.

8.3.7 The RNA world

Rich (1962), Woese (1967), Orgel (1968, 1986), and Crick (1968) suggested that: "Possibly the first enzyme was an RNA molecule with replicase properties." The discovery of *ribozymes*, in which RNA plays both roles, namely, that of recording and transfer of information and the catalysis role of enzymes proved this prophetic (Cech, 1986; Zaug and Cech, 1986). This proposal is now known as *The RNA world* as coined by Gilbert (1986). Some thought ribozymes provided the pathway from the *Urschleim* to the protobiont. It was proposed as a self-replicating system preceding the present system, which did not depend on the enzyme action of proteins but, rather, on a pre-enzyme action of nucleic acids (Joyce, 1998; 2002a, 2002b; Szostak, Bartel, and Luisi, 2001). This suggestion seems to move the problem a step nearer to the protobiont but still encounters the primary questions of the handedness of amino acids and ribose sugars, the generation of the genetic message, and the origin and evolution of the genetic code.

Taken at best, however, this proposal simply moved the search for an origin of life scenario one step from the *Urschleim*, where it encounters the need for a chemical procedure for the prebiotic formation of the D form of the sugar ribose, a component of RNA (Joyce, 1991). There is only one plausible synthesis of ribose that may be considered in the prebiotic milieu, namely, the polymerization of formaldehyde. Ribose is only one of a number of sugars and never the primary product (Orgel, 1986). This can hardly be called a prebiotic synthesis, because it requires a well-educated chemist acting as *deus ex machina*. Furthermore, the condensation of adenine or guanine with ribose leads to mixtures of the optical isomers that are very complex. From the point of view of the biochemist, this problem is as difficult as the one the discovery of ribozymes was supposed to solve. Furthermore, nonbiological reactions in the prebiotic milieu lead to $2'$, $5'$ isomers (Orgel, 1986). This does not lead in the direction of the origin of life, because the reactions that are typical of biochemical products are almost always the $3'$, $5'$ linked isomers, whether or not they are catalyzed by protein enzymes. RNA is hydrolyzed about one hundred times more rapidly than DNA and would not appear in quantity in the *Urschleim*. Furthermore, this proposal is not as promising as seemed at first, as RNA is not as versatile a catalyst as protein enzymes. Moreover, RNA is itself a molecule more complex than amino acids. Furthermore, Nelson, Levy, and Miller (2000) report that:

> Numerous problems exist with the current thinking of RNA as the first genetic material. No plausible prebiotic processes have yet been demonstrated to produce the nucleosides or nucleotides or for efficient two-way nonenzymatic replication.

Even those who believe in an *Urschleim* do not regard it plausible that RNA existed in large quantities, as though that decision were up to them. The nucleotide cytosine is a component of DNA and mRNA. Shapiro (1999) has examined the question of the prebiotic synthesis of cytosine. He found:

> No reactions have been described thus far that would produce cytosine, even in a specialized local setting, at a rate sufficient to compensate for its decomposition. On the basis of this evidence, it appears quite unlikely that cytosine played a role in the origin of life.

Even more striking is the paper by Levy and Miller (1998), who found that at 100°C the half lives of A and G are about one year; that of U, twelve years; and C, nineteen days.

8.4 Haeckel's *Urschleim* and the current speculations on the origin of life

Bada and Lazcano (2002a 2000b, 2003), in spite of the facts discussed in this chapter and elsewhere in this book, describe what they represent as the modern conjectures of the origin of life that would do well for a Kipling's *Just So Story*. They pick and choose the processes that they propose led to life from Haeckel's *Urschleim*.

> If the transition from abiotic chemistry on the early Earth indeed took place at low temperatures, it could have occurred during cold quiescent periods between large sterilizing events.

Although there were "large sterilizing events," there were no cold quiescent periods.

The citric acid cycle is the foundation of intermediary metabolism in autotrophs (Morowitz et al., 2000). In some autotrophs, CO_2 is taken in and the molecules of the cycle are synthesized. This proposal for the origin of life is like a recipe for rabbit stew: First catch the rabbit! It is no longer believed that the atmosphere of the early Earth was reducing, at the time when life emerged, before 3,850 million years ago (Kasting, 1993; Mojzisis, Krishnamuthy, and Arrhenius, 1999). The modern view is that the early atmosphere, in the era between 4.0×10^9 and 3.850×10^9 years, was neutral and composed of N_2, CO_2, H_2O, with perhaps some NH_3 (Canuto et al., 1983; Hart, 1978; Walker, 1976). Time, as it often does, has rendered obsolete the autogeny of Haeckel (1866, 1905), the coacervates of Oparin and Bungenburg de Jong (1932), and the hot dilute soup of Haldane (1929).

There is a need for make-believe (Hoffer, 1951) to support the parable of the *Urschleim*. It is widely regarded by the authors of papers on the origin of life, that they are contributing to the glory of the essential tasks of science, being so well established that no justification is necessary (Bada and Lazcano, 2002; Doudna and Cech, 2000; Joyce, 2002a, 2000b; Schopf, 1999). Engel's (1954) comment is the basis of this make-belief: "Life is the existence of protein bodies." This Marxist canard is taught as scientific fact in college text books (Watson et al., 1987) and in popular books on science (Chaisson, 2001; de Duve, 1991, 1995; Schopf, 1999; Wills and Bada, 2000). Once this canard gets into textbooks and the popular science literature, it stays there for generations.

One would suppose that the workers in this field would have given a first priority for the establishment of the existence of the *Urschleim*. Miller,

Schopf, and Lazcano (1997) admit that there is no geological or geochemical evidence that an *Urschleim* ever existed. Strange to say, they seem to have left this for later in the manner of an ingenious architect, as reported by Captain Lemuel Gulliver in Jonathan Swift's *Gulliver's Travels*. This architect contrived a new method for building houses by starting at the roof and working down and establishing the foundation at the end of the project. The architect pointed out that among the obvious advantages of this method is that once the roof was in place the workers could toil in the shade of the hot sun and at other times be protected from rain and snow. Thus, inclement weather would not delay the progress of the construction. The Grand Academy of Lagado had approved this proposal by peer review, but the architect had not yet put it into practice at the time of Captain Gulliver's visit.

8.5 The Central Dogma and the origin of life

As I mentioned in Chapter 3, no code exists to send information from protein sequences to mRNA or DNA (Battail, 2001; Billingsley, 1965, 1995; Kolmogorov, 1958; Ornstein, 1974; Shields, 1973; Yockey, 1974, 1978, 1992, 1995, 2000a, 2002a). Therefore, it is *impossible* that the origin of life was "proteins first" from the *Urschleim* (Section 3.1.1). Nevertheless, "proteins first" is widely taught in university classrooms (Schopf, 1999) and perhaps at the Grand Academy of Lagado as well. The funding of origin of life speculations in the Astrobiology programs supported by NASA (Morrison, 2001; http://astrobiology.arc.nasa.gov) and other agencies, based on discovering means of generating "proteins first" in the prebiotic ocean, must be completely reorganized.

A good experiment is always a good experiment. Unfortunately, the interpretation of the corpus of publications on the origin of life is false. Those experiments are based on a belief that life is just complicated chemistry and that the pathway to the origin of life, if it could be found, is emergent from organic chemistry.

8.5.1 Old scientific fables never die! they don't even fade away!

The fable about the origin of life from *protoplasmal primordial atomic globules* in a racemic *Urschleim* can not be killed with facts, does not fade away, and appears to be immortal! The Scripps Research Institute celebrated the fiftieth anniversary of Miller's "land mark experiment" [*sic*] on June 10,

2003, with a lecture by Professor Bada (Press Release, Monday May 19, 2003).

Francis Crick (1981), rather dryly, asked:

> As far as I know, no one, rather surprisingly, has deliberately tried to grow bacteria in an artificial "soup" made in a Urey–Miller type of experiment (most of the experimenters go to great lengths to exclude microorganisms from their incubation flasks), but one would certainly expect many types of bacteria to thrive there, even in the absence of atmospheric oxygen.

The editors of *Nature* published a review (Heilbron and Bynam, 2002) of significant events since 1551 A.D., which included Miller's paper (1953). *Nature* published my note (Yockey, 2002b) reminding them that Miller had merely repeated the work of Walther Löb (1913) and Oskar Baudisch (1913). Haldane had refereed to the work of Baly et al. (1922). Bada and Lazcano (2002b) replied at last recognizing Löb's work and with a long discussion of what they believe is the significance of Miller's work. *Science* published a paper by Bada and Lazcano (2002b) extolling the "pioneering work" of Oparin and Haldane and the prebiotic soup speculation. I replied (Yockey, 2003) with the remarks in Section 8.2.3. Bada and Lazcano (2003) returned to *Science* with the same arguments supporting the prebiotic soup speculation. But see my reply (Yockey, 2003).

We have rounded up the usual suspects and the status of research on the origin of life is still, *Omne vivum ex vivo*. There is no yellow brick road from Haeckel's *Urschleim* to the origin of life. It is astonishing to some people that some things are impossible or unknowable. I shall take this up in Chapter 11.

Of course, none of the speculations on the origin of life discussed earlier addresses the critical issues in molecular biology, which are the subject matter in this book, namely, the origin and function of the genetic code and the primitive genome.

9

Philosophical approaches to the origin of life

The standard model of the origin of life on Earth was first given form by Oparin (Oparin 1957) and Haldane and begins with the abiotic production of organic molecules. Abiotic chemical evolution is then thought to occur in the presence of liquid water, resulting in more complex structures leading to life itself and the onset of abiological evolution. The standard model for the origin of life posits that life is a naturally emergent property of matter in Earth-like environments and would develop rapidly on any similar body.

Christopher P. McKay in his ISSOL *Urey Prize Lecture at NASA: Planetary Evolution and the Origin of Life* (1989). NASA Space Sciences Division, Urey Prize Lecture, also in *Icarus*, **91**, 93–100 (1991)

9.1 Vitalism

Vitalism is the belief that there is a metaphysical, supernatural, nonmaterial, idealist *élan vital*, a life force that distinguishes living from nonliving matter. Vitalism has its roots in the German idealist philosophy of Georg Wilhelm Friedrich Hegel (1770–1831), F.W.A. Schelling (1775–1854), and L. Oken (1779–1851) in the nineteenth century, members of a romantic philosophic movement, *Naturphilosophie*, who believed all creation was a manifestation of a *World Spirit*. They believed all matter possessed this *Spirit* and organized bodies had it to an intense degree. In the nineteenth century, it was quite possible to be a vitalist, believing in a vital force or *élan vital*, without thinking of the vital force being supernatural. At the time it was as valid to attribute the laws and effects of vitality to a nonmaterial vital force as it was to attribute the laws and effects of gravity to a nonmaterial gravitational force.

According to Troland (1914):

> The difficulties that the neo-vitalists urge against the mechanistic theory of heredity may perhaps be summarized as follows. In the first place the germ-cell, which must be regarded as of prime importance in any mechanistic view of heredity, is said to be too small to contain a physical machine that can be capable of determining with accuracy all of the intricacies of structure and function which are exhibited in higher organism; in the second place no definite conception has been advanced to show how such a hypothetical determining influence can exist.

(We know now that DNA, a mechanistic source of heredity, can indeed determine with accuracy a message that determines much of the intricacies of structure and function of all life.)

Friedrich Wöhler (1800–82) produced urea, a biochemical substance that comes only from organisms, from ammonium cyanate, an inorganic chemical compound (Wöhler, 1828). As Wöhler wrote in his letter to his teacher, Berzelius (1901): "... and I must tell you that I can prepare urea without requiring the kidneys of an animal, either man or dog; the ammonium salt of cyanic acid is urea." Some denied that urea was an organic compound but was, rather, merely a waste product, such as water. It is now known that urea and ammonium cyanate are an example of isomerism; they share the same empirical formula.

Although today we see Wöhler's work as a stake in the heart of vitalism, it persisted into the first half of the twentieth century. It was kept alive by the chemist-philosopher Hans Driesch (1914) and the philosopher Bergson (1859–1941) who was awarded the Nobel Prize in Literature in 1927. Bergson (1944) believed that vitalism was required to understand the apparent violation of the second law of thermodynamics by the increase of complexity in evolution.

Vitalism no longer plays a role in biology (Mayr, 1982), but it left a legacy of "organic chemistry" contrasted with "inorganic chemistry" and the need to be specific about "biochemistry." Thus, urea is a biochemical and an organic chemical. Its isomer, ammonium cyanate, is inorganic.

9.2 Mechanist-reductionism

The idealist philosopher Immanuel Kant (1724–1804) anticipated the views of Charles Darwin in his *Critique of Judgment* (1790):

> This analogy of forms, which in all their differences seem to be produced in accordance with common prototype, strengthens the suspicion of a real kinship by descent from a common primordial ancestor (*Urmutter*). This we might trace in the gradual approximation one animal genus to another, from that in which the principle of ends seems best authenticated, namely from man, back to the polyp (jelly fish), and finally to the lowest perceivable stage of nature, to crude matter: from this and from the force within, by mechanical laws (like those by which it acts in the formation of crystals), seems to be derived the whole technic of nature which,

in the case of organized beings, is to us that we feel compelled to imagine a different principle. (This quotation in English translation can be found in Wächtershäuser, 1997.)

A group of young physiologists at an 1869 meeting in Innsbruck, Austria, declared:

> The ultimate objective of the natural sciences is to reduce all processes in Nature to the movements that underlie them and to find their driving forces, that is, to reduce them to mechanics. (Mayr, 1982)

Rudolf Virchow (1821–1902), a very famous physician in the nineteenth century, in 1858 objected to *Naturphilosophie* (Rather, 1958):

> As particular, as peculiar, and as much interiorized as life is, therefore so little is it withdrawn from the rule of chemical and physical law. Rather does every new step on the path of knowledge lead us near to an understanding of the chemical and physical processes on whose course life rests.
> Natural law is the miracle, and this law fulfills itself in a mechanistic manner on the path of causality and necessity.
> But if there was an origin of life, then it must be possible for science to fathom the conditions of this origin. At present this is an unsolved problem.

9.3 Dialectical materialism

The dialectical materialism of Friedrich Engels (1820–95) was the foundation of Leninism–Stalinism. According to dialectical materialism, the appearance of life is achieved not through the laws of physics and chemistry but through the *Law of the Transformation of Quantity into Quality*. The transformation of quantity into quality precluded the possibility of explaining the origin of life and life processes in elementary physicochemical terms. Engels considered Darwin's theory of evolution to be an example of the transformation of quantity into quality (Graham, 1972, 1987, 1993; Oparin, 1957).

Friedrich Engels began *The Dialectics of Nature* about 1875, in which he established the nature and origin of life speculation for dialectical materialism:

Life is the existence of protein bodies, the essential element of which consists in *continual metabolic interchange with the natural environment outside them*, and which ceases with the cessation of this metabolism, bringing about the decomposition of the protein. If success is ever attained in the preparation of protein bodies chemically, they will undoubtedly exhibit the phenomena of life and carry out the mechanisms of metabolism, however weak and short-lived they may be. (italics in the original)

Engels spent his last years editing the works of Karl Marx and did not live to finish *The Dialectics of Nature*. The document was found in his papers and published in Moscow in 1925 in the original German and in a Russian translation.

Engels had observed in one of the fragments of *The Dialectics of Nature* (Engels, 1954):

Mechanism applied to life is a hopeless category, at the most we could speak of chemism, if we do not want to renounce all understanding of names.

The proponents of dialectical materialism are extremely hostile to mechanist-reductionism. Alexandr Ivanovich Oparin (1894–1980), on reading that quotation, abandoned the reductionism-mechanism of his first paper (Oparin, 1924), and became an immediate convert to dialectical materialism. Oparin (1957) went into great detail to denounce reductionism-mechanism and to support dialectical materialism.

Western men of words (Hoffer, 1951), intellectual apologists for Oparin, who are not trained in the intricacies of philosophy may find the relationship between reductionism-mechanism and dialectical materialism to be a distinction without a difference (Lazcano, 1997; Miller, Schopf, and Lazcano, 1997; Schopf, 1999; Wills and Bada, 2000).

9.4 N. I. Vavilov, H. J. Muller, T. D. Lysenko, and A. I. Oparin

The Soviet Union party line could not accept "Morgano-Weissmannite genetics" based reductionism-mechanism. According to that heresy, the upper classes of the czarist regime were therefore intrinsically superior to the proletariat. According to the party line, total control of the environment by the party would allow communism to create the New Soviet Man *immediately.* Woe betide anyone who suggested that the New Soviet Man would not emerge simply by changing his environment!

The distinguished Russian geneticist Dr. Nikolai Ivanovich Vavilov (1887–1943), at one time president of the Lenin All-Union Academy of Agriculturist Science in the Soviet Union, had collected a large number of seeds, fruits, and tubers from parts of Transcaucasia, Ethiopia, and Afghanistan, which are resistant to rust and mildew. This collection and others from all over the world was preserved at the *Vavilov Institute of Plant Industry* in Leningrad. During the 880-day siege by the Third Reich during World War II, tens of thousands of the city's residents starved to death. Eight of Vavilov's scientists died of starvation, although they could have eaten the seed collections that were readily at hand. These collections were preserved and guarded because the people's agricultural future depended on the diverse gene stocks (Davis, 2003; Webster, 2003).

The distinguished Russian geneticist Dr. Nikolai Ivanovich Vavilov (1887–1943) invited the American Professor Herman J. Muller (1890–1967) to head a genetics laboratory in Moscow. Muller was a lifelong Marxist in spite of an unhappy stay in the U.S.S.R. Muller was, in his Moscow period, to run afoul of Trofim Desinovich Lysenko (1898–1967) and Alexandr Ivanovich Oparin. Muller (1966) referred bitterly to Oparin as a Lysenkoist who subtly carried out an attempt to downgrade the significance of genetics. Muller had been elected a foreign member of the Lenin All-Union Academy of Agriculturist Science. When he learned the fate of many of his Soviet friends, he resigned his membership in 1948 and returned to the United States. The U.S.S.R. Academy of Sciences accepted this heretical action without regret for " . . . openly joining the enemies of progress and science of peace and democracy" (Medvedev, 1969).

Stalin regarded Vavilov as one of the "bourgeois specialists" of the Czarist regime. Among Vavilov's ideological crimes [*sic*] were his scientific work in the United States and, most seriously, his rejection of dialectical materialism. Because of that, some 150 of the scientists in the *Vavilov Institute of Plant Industry* were arrested and executed or sent to Stalin's GULAG (Burch, 2003). Vavilov attacked Lysenko and his foolish notions about genetics during the great purges of 1937–38. Lysenko denounced Vavilov, who was arrested on August 6, 1940.

Oparin was very close to Lysenko. His summer dacha adjoined Lysenko's on the outskirts of Moscow (Schopf, 1999). His complicity in the denunciation and death of Vavilov can not be overlooked. Oparin was well fed, as Schopf (1999), who knew him, pointed out. Vavilov, by contrast, died of starvation in Stalin's torture chambers at Saratov prison on January 26, 1943 (Soyfer, 1989, 1994).

Western True Believers, intellectual apologists for Oparin, do not weep for Vavilov (Lazcano, 1995; Miller, Schopf, and Lazcano, 1997; Schopf, 1999; Wills and Bada, 2000). The *Vavilov Institute of Plant Industry* in St. Petersburg continues its heroic work supporting Russian agriculture, now fully devoid of Lysenkoism (Davis, 2003).

By the miracle of the Internet, I have been corresponding with scientists who were in the Soviet Union at the time of Lysenko and Oparin. They tell me that I am absolutely correct in my description of Oparin's special role in Stalin's crushing of Soviet genetics and the destiny of Dr. Nikolai Vavilov personally. After the death of Stalin, the national headquarters of the Soviet Academy of Sciences was located on Vavilov Street (Jukes, 1997). My correspondents in Russia tell me that there is now, unfortunately, also an Oparin street in Moscow.

9.5 Lysenko, Oparin, and the Heresy of "Morgano–Weismannite genetics"

Because of his peasant background, Stalin (1879–1953) regarded Lysenko as the ideal man to take over the work of Michurin, an uneducated plan breeder:

1. To replace the "bourgeois specialists" of the Czarist regime, almost all of whom had been arrested by the late 1930s;
2. To vindicate the New Soviet Man; and
3. To make Nature serve the people (Medvedev, 1969; Soyfer, 1989).

In 1937 Stalin made Lysenko a member of the Supreme Soviet and head of the Institute of Genetics of the Soviet Academy of Sciences, Lysenko applied his spurious theory of horticulture to remake Nature. He became the autocratic boss in biology and agriculture in the U.S.S.R., where American author Lincoln Steffens (1866–1936) had seen the future and found that it worked.

The All-Union Academy of Agricultural Sciences was under Lysenko's leadership during its session in August 1948, which destroyed genetics for more than a generation in the U.S.S.R. Stalin had appointed thirty-five new members to the Academy, whose names were published in *Pravda* on July 28, 1948, to assure the vote would go according to his instructions. The Presidium of the U.S.S.R. Academy of Sciences passed resolutions that called for the removal from the Scientists' Council those who supported

"Morgano–Weissmannite genetics" and stated that "Michurin's materialist direction in biology is the only acceptable form of science, because it is based on dialectical materialism and on the principle of changing Nature for the benefit of the people." Immediately after that session, about three thousand scientists were dismissed from their posts. Russian genetics ceased to exist (Soyfer, 1989).

Oparin was the *only* well-known biologist to join the Lysenkoist movement. From 1948 to 1955, he was in charge of the Lysenkoist hiring and firing within the Academy of Science (Joravsky, 1970). Wills and Bada (2000), apologists for Oparin, admit that: "There is no doubt that he acquiesced in, and may even have been directly involved in, the exile of a number of scientists to GULAG."

Those who knew their way around the Soviet system could still get their views published. The Soviet astronomer I. S. Shklovsky (1916–85) wrote a book entitled *Universe, Life and Mind* published in the U.S.S.R. in 1961 in which he went far beyond astronomical topics with the intention of demolishing Oparins's *notorious theory*, as he put it. This was dangerous because Lysenko was still in favor in 1961, and Oparin was his close confederate. In a later book, *Five Billion Vodka Bottles to the Moon*, Shklovsky wrote of the reaction to his assault on Oparin, "To all intents and purposes the book had escaped censorship. There was an uproar over it but nothing terrible happened. Oparin squealed in indignation. I wrote him a polite letter, but he ripped it into little pieces and returned it in the same envelope."

Shklovsky demolished Oparin's "notorious theory" in *Intelligent Life in the Universe* a book he coauthored with Carl Sagan. Shklovsky writes, using the editorial *we*:

> We find it difficult to agree with Oparin that these coacervate droplets were the first forms of life on Earth. While the analogy between material exchange and metabolism is interesting, it hardly proves that coacervates were primitive living organisms. A fundamental property of living systems is self-replication, including the presence of a genetic code which transfers properties from generation to generation. Coacervates have no mechanism of inheritance. Oparin's hypothesis does not explain the transition from a nonliving to living system. p. 241)

Shklovsky is committing the heresy of denying the dialectical materialist *Law of the Transformation of Quantity into Quality* and its manifestation as

Michurinism–Lysenkoism (i.e., Nurture rules Nature). Shklovsky commits a further heresy by supporting the modern science of genetics, including the discovery of the genetic code. Shklovsky's objection to the coacervate proposal is the same as that made by Jacques Loeb in 1906, 1912, 1916 and 1924 before the discovery of the genetic code.

Oparin's own writing show he contributed to making science the slave of ideology and social policy. They are a rejection of the entire science of genetics ninety-seven years after the publication of Gregor Mendel's "Experiments on plant hybrids," which started the modern science of genetics.

Apologists for Oparin among the Western intelligentsia (Schopf, 1999; Wills and Bada, 2000), claim that he chose not to follow the fate of Vavilov, Boris Numerov (1891–1941), and many others, because he was simply submitting to the pressure to conform to the party line and the very strong instinct for survival. In the opinion of people who knew him, Muller (1966), Shklovsky (1991), and Zhores A. Medvedev (1969), in particular, Oparin gave Lysenkoism much more support than was necessary and for much longer than was necessary. Oparin shows his devotion to Lysenkoism did not spring from fear of Stalin by his reverent quotations of Lysenko in his book, *LIFE: Its Nature, Origin, and Development*, published in 1964, *eleven years after Stalin's death*:

> In the course of its development each contemporary plant must pass through certain definite stages which are necessary parts of its onotogenesis and which follow one another in a strict, orderly sequence, each stage requiring its own particular set of environmental conditions for its completion. According to T. Lysenko these stages are, in essence, qualitatively discontinuous changes in metabolism and form the basis of the 'readiness' of plants for the occurrence (under suitable environmental conditions) of perceptible phases of morpho-genesis (changes of tillering, of flowering, etc.) (pp. 177–8).

9.6 The defrocking of Alexandr Ivanovich Oparin

As Oparin himself pointed out frequently, he got most of his ideas from Friedrich Engels's *The Dialectics of Nature* (1954). He acknowledged (Oparin, 1957) that Haeckel had proposed the monistic hypothesis of abiogenesis in 1866 and proposed that the nucleus of the cell carried inheritance. Oparin was far from being the first to propose a primeval soup in the early ocean as his apologists have claimed (Wills and Bada, 2000; Schopf, 1999).

Western intellectuals, men of words (Bada and Lazcano, 2000a, 2000b, 2003; Lahav, 1999; Lazcano, 1997; Miller, Schopf, and Lazcano, 1997; Schopf, 1999; Wills and Bada, 2000) have a strange proclivity to fool themselves and, unfortunately, their students. They have often suggested that Oparin made a major breakthrough by his coacervate model of living cells. (As I show in Chapter 8, this idea is due to Haeckel in 1866.) J. William Schopf referred to Oparin as "the great man." On the contrary, Oparin (1957) quoted extensively the long paper, *"Die Koazervation und ihre Bedeutung für die Biologie"* (*Coacervation and its meaning for Biology*), by H. G. Bungenberg de Jong (1932). Although Jacques Loeb (1924) was a very famous scientist and an expert in coacervates, he was not cited in Bungenberg de Jong's paper.

When one reads the literature it becomes quite clear that Oparin made no contribution to the origin of life or to molecular biology (Jukes 1996 and 1997). If anyone were to receive credit for Pooh-Bah's *"protoplasmal primordial atomic globule"* or coacervate proposal, if not Haeckel or Engels, it should be Bungenberg de Jong (1893–1977).

This is consistent with the dialectical materialism of Friedrich Engels whose philosophy, along with that of Karl Marx, was the foundation of Leninism-Stalinism. It is curious that Western men of words are sympathetic to the Communist Party and are morally blind to the crimes of Stalin (Amis 2002; Hoffer, 1951). This is in spite of the works of Aleksandr Solzhenitsyn in *The GULAG Archipelago* and other writings (Applebaum, 2003).

These facts do not discourage faith in the official NASA policy for the origin of life (Morrison, 2001). That policy is based on the Marxist dialectical materialism of Engels and Oparin. NASA policy follows Engels "proteins first" from dialectictical materialism. In Chapter 3, I noted that "proteins first" speculations are impossible, according to the Central Dogma.

10

The error catastrophe and the hypercycles of Eigen and Schuster

A little learning is a dangerous thing;
Drink deep, or taste not the Pierian spring:
There shallow draughts largely intoxicate the brain,
And drinking largely sobers us again.
A little learning

Alexander Pope (1688–1744)

10 Introduction

It often happens that two academic schools of thought, dealing with the same subject, totally ignore each other. "Can we use information theory to solve our problem of self-instruction?" (Eigen, 1971). Eigen and his colleagues use some theorems from information theory, but they have neglected the more important ones so that their work is not based on the axioms of information theory (Eigen, 1971, 1992, 2002). The most important ones they neglected are the Channel Capacity Theorem and the Shannon–McMillan–Breiman Theorem. The Göttingen School and Professor Dr. Manfred Eigen have made many important contributions to science. They are, however, not above comment.

10.1 Eigen's remedies for what he sees as inadequacy in Shannon's information theory and the Shannon–McMillan–Breiman Theorem

Shannon (1948), in his second paragraph, warned against confusing the semantic aspects of a message with the properties of the communication system that apply to any message. Eigen (1971) regards that as an inadequacy of classical information theory.

> Eigen (1971) states that: "The information resulting from evolution is a *valued information* and the number of bits will not tell us too much about its functional significance." ... "If entropy

characterizes the amount of "unknowledge," then any decrease of "unknowledge" is equivalent to an increase of "knowledge" or "information." The complementarity between information and entropy shows clearly the limited application classical information theory has to problems of evolution. Whenever information has a defined meaning, e.g., as in language by agreement, or in biology after evolution has brought the fixation of a code.

Of course Eigen's comment applies to any message.

He states that:

> Such sequences cannot yet contain any appreciable amount of information.

This is merely a play on words by using *information* in the sense of *knowledge, meaning* and *specificity*. Like Watson and Crick, he confuses the genetic message with the genetic code. To the computer-literate reader, all sequences, that is, bit strings in their hard drives, have a measurable "amount of information," although they do not necessarily have *knowledge, meaning* and *specificity* (Section 1.2.1).

Those authors who put information in quotes are impling that they don't understand the genetic message. Shannon entropy is a measure of uncertainty, randomness, ignorance, order and disorder. Shannon entropy, in Equation 7, Chapter 4, controls the number of sequences with which we need to be concerned. Contrary to Eigen (2002), the most obvious parameter is not the error rate but, rather, the Shannon entropy.

Eigen proposes a "master sequence" in a "quasi-species" each one having a "value parameter" or "superiority parameter" (Eigen, 1971, 1992, 2002; Eigen and Schuster, 1977, 1978a, 1978b; Eigen, Winkler-Oswatitsch, and Dress, 1988). It is misleading to believe that there are only a few "master copies," "master sequences," or "consensus sequences" in homologous proteins. I made this point in (Yockey, 1992, pp. 141 and 274) and it is important to emphasize it again. Unfortunately, the "quasi-species," "masters sequence," "value parameter," and "superiority parameter" notions of Eigen and Schuster seem to be well established in the literature and will be very difficult to remove.

The genetic system is very nearly the same in all organisms and uses the same nucleotides and amino acids. The genetic information system, like all communication systems, operates without regard for the specificity, or value, of the message. It must be capable of handling *all* genetic messages of *all* organisms, extinct and living, as well as those not yet evolved. The genetic signal or message does not have to be "about something."

Eigen (1971, 1992, 2002), Eigen et al. (1988), Eigen and Schuster (1982), and Schuster (2000) calculate the number of natural peptide sequences as 20^v and the number of polynucleotide sequences as 4^v where v is the number of sites in the sequence under consideration. As I have shown in Chapter 4, Chapter 6, and elsewhere (Yockey, 1974, 1977, 1981, 2000, 2002), this calculation includes an enormous number of "junk sequences" and that invalidates these papers. The Shannon–McMillan–Breiman Theorem (Section 4.1) tells us that the number of sequences in the "all residues" set is 2^{NH} and they are all nearly equally probable. From Table 6.3, there are $6.42392495176 \times 10^{111}$ functionally equivalent "master sequences" in the "all residues" set for 1-iso-cytochrome c. If one makes this calculation using Expression 1, the result is:

$$20^{113} = 1.03845927171 \times 10^{147}. \tag{10.1}$$

Thus, doing the problem correctly, one finds that the 1-iso-cytochrome sequences are only a very tiny fraction $6.18601471259 \times 10^{-36}$ of the total possible sequences (Section 6.4). This result comes directly from the elementary theorems of probability theory where we encountered the notion of Shannon entropy and a measure of "uncertainty."

Making the calculation correctly, as Shannon (1948) did, leads to the Shannon–McMillan–Breiman Theorem and an understanding of Shannon entropy as a measure of "uncertainty." Therefore, Eigen and many others grossly overestimate the number of sequences that need to be considered (Section 6.4).

10.2 The hypercycle and views of the Göttingen School on the origin of life

The work of Eigen and his colleagues is based on the primeval soup paradigm (Section 8.4). I don't find a reference to Haeckle's racemic *Urschleim* (Ch. 8) or the role of the Central Dogma that prohibits "proteins first" (Yockey, 2002, Ch. 3). Eigen does not mention the paradox that enzymes are required to define or generate the reaction network, and the network is required to synthesize the enzymes and their component amino acids.

The essence of the origin of life scenario by Eigen and Schuster (1977, 1979, 1982) is the "hypercycle." They state that (1978b) no less than 10^8 *identical copies* must be present to start the hypercycle. I have described in Section 8.1.3 the Guru's triumph of Mind over Matter. Even a short sequence of one hundred outcomes of the coin toss leads to a total of $2^{100} = 1.2676506002 \times 10^{30}$ sequences. The idea that after the first toss, there could be an *identical* copy in a second or any subsequent toss within the age of the Earth is statistically beyond belief.

> The hypercyclic link would then become effective only after concentrations have risen to sufficiently high level. There is only one solution to this problem: The hypercycle must have a precursor, present in high natural abundance, from which it originates gradually by a mechanism of mutation and selection.

The Göttingen School does not mention that the *Urschleim* was necessarily racemic, and therefor could not be a source of protein or nucleic acids (Pasteur, 1848). One of the more important objections to the hypercycle paradigm is that there is no evidence that the *Urschleim* was dense enough for the hypercycle to start (Section 8.4).

He rejects the origin of life by chance scenario although he says that the origin of life must have started by self-organization from random events in a molecular chaos without functional organization (Eigen, 1971).

> At the "beginning," whatever the precise meaning of this may be – there must have been *molecular chaos* without any functional organization among the immense variety of chemical species. Thus the self-organization of matter we associate with "the origin of life" must have started with random events. (Eigen, 1971)

He defines "random" as:

> The term "random," of course, refers to the non-existence of functional organization and not to the absence of physical (i.e., atomic, molecular or even supramolecular structures.)

The term "random" lies in the field of the mathematicians. This is hardly the definition of "random" due to Kolmogorov (1958, 1965), Martin-Löf (1966), and Chaitin (1966, 1975a, 1975b, 1979, 1987b, 2001), who published their work on computational complexity and randomness culminating in their algorithmic information theory. I have given a comparison of "randomness" and "complexity" as applied in molecular biology in Chapter 11.

10.3 Eigen's error catastrophe

The term "error catastrophe," introduced originally in Eigen's theory of evolution (Eigen, 1971; Eigen and Schuster, 1982), has unfortunately, become fashionable and misleading in virology (Croty, Cameron, and Andino, 2001; Eigen, 2002; Grande-Pérez et al., 2002). Eigen (1971) and Eigen and Schuster (1977, 1978a, 1978b, 1982) address the question of the effect of errors and they arrive at the following equation:

$$v_{\max} = \frac{\log_e s_m}{(1 - \overline{q}m)}, \tag{10.2}$$

where s_m is the "superiority parameter" for the sequence, $v_{\max}$ is the maximum length of a sequence in which the expectation of error is below a geometrical average quality factor $\overline{q}_m$. Eigen and Schuster (1977, 1978a, 1978b, 1982), put forward Equation 10.2 as defining a threshold at which an "error catastrophe" will occur: "If this limit (i.e., $v_{\max}$) is surpassed, the order, i.e., the equivalent of information, will fade away during successive reproductions." Here again Eigen confuses "order" with "information."

I am not the only one to have noticed that Eigen's treatment of error in the genetic system predicts that life would not survive very long (Davis, 1999; Smith and Szathmáry, 1995, 1999; Yockey, 1992). This a fatal objection to Eigen's theory of "error catastrophe." Unfortunately, these authors (Davis, 1999; Smith and Szathmáry, 1995, 1999) accept Eigen's "error catastrophe" and try to find a way around it.

This misunderstanding could have been avoided if it were not for the *apartheid* in the olive groves of academe. Shannon's Channel Capacity Theorem (Shannon, 1948), had firmly established the conditions by which messages may be communicated with minimum error. Shannon (1948) showed that the proper way to compare two or more sequences is by the effect of error accumulation on mutual entropy (Yockey, 1992, 2000, 2002). Contrary to Eigen (2002), the effect of error is not a sharply defined limit; rather, the information content in the genome is reduced gradually (Yockey, 1958, 1992, 2000, 2002).

Life tables do not show an "error catastrophe." On the contrary, they show a ledge followed by a gradual decline (Hekimi and Guarente, 2003; Lee, Tolonen, and Ruvkun, 2003; Murphy et al., 2003). In the case of humans one has the Gompertz function (Gompertz, 1825) that shows a sharp drop due to infant mortality and then a long gradual decline to old age. The same result has been shown by Bongaarts and Feeney (2003) in their study of Swedish females born in 1850. The proper treatment of the question of "error catastrophe" is taken up in Chapter 5.

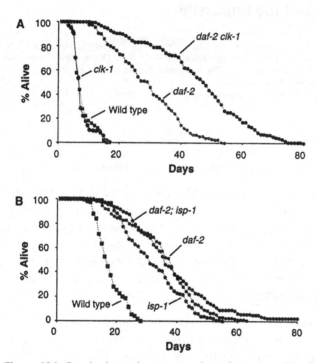

Figure 10.1. Genetic interactions among longevity genes. (A) *daf-2(e1370) clk-1(e2519)* double mutants live a very long time, indicating that almost all age-dependent degenerative processes in the worm can be substantially prevented by altering only very few genes at a time. Animals were raised at 20°C and then permanently placed at 25°C. Adult life-span is shown. The data are from (2). (B) *daf-2(e1370)* and *isp-1(qm150)* increase life-span to a similar degree but do not show a positive interaction, because the double mutants *dat-2; isp-1* do not live appreciably longer than the single mutants. Adult life-span at 20°C is shown. The data are from Hekimi and Guarente (2003).

11

Randomness, complexity, the unknowable, and the impossible

"It's no use trying," said Alice. "One can't believe impossible things." "I dare say you haven't had enough experience," said the White Queen. "When I was your age, I did it for an hour a day. Why sometimes, I've believed as many as six impossible things before breakfast."
Alice in Wonderland, Lewis Carroll (1832–1898), aka Reverend Charles Lutwidge Dodgson

11.1 Orderliness, randomness, and self-organization

There has long been considerable opinion in the literature that life must be nonrandom and appears because of self-organization (Eigen, 1971). Many authors appeal to pattern formations that abound in Nature, such as crystal structure, stripes on a zebra, snowflakes, oil drops in water, hurricanes, and ripples on a river bed. It is often said that the crucial event in the origin of life is the saltation from a disordered state to an ordered state (Chaisson, 2001; Dyson, 1982; Schrödinger, 1987, 1992).

> What has really happened, if we compare the two selected species after the restoration of the steady states, is a change in "valued information" which is reflected in an increased order. (Eigen, 1971)

> Thus, combinatorial explosion is a universal threat to biopolymer sequences and structures as well as reaction and controlling networks. Examples are known from biology, in particular metabolic, genetic, developmental, signaling, and neural networks. If this is true, how then, can organized objects originate? Must not all processes that are not regulated externally end up in a highly diverse mess of molecular species, each one at best realized in a few molecules? The frequently given answers invoke self-organization as a (universal) principle introducing order into diverse manifolds. (Schuster, 2000)

Accordingly, the thrust of many origin of life scenarios has been to attempt to show how to generate "order" out of "chaos" (Morowitz et al., 2000). Those who pursue this approach are caught up by the Tar Baby, like Br'er Rabbit, and get into more and more trouble (*The Complete Tales of Uncle Remus*, Joel Chandler Harris, 1955). Appeals to "self-organization" are a misunderstanding of the genetic message in the genome discussed in Section 1.1.2.

Randomness is often regarded as a vague "buzzword" concept that cannot be measured. One often finds scientists using expressions such as "selected at random," "randomization," "nonrandom," expressed without a clear and quantitative definition (See Section MA2–1.2, Random Variables). Such people (Chaisson 2001; Morowitz et al., 2000; Orgel, 2000; Schuster, 2000) have not been listening. Consistent with the approach in this book, I shall discuss how these concepts can be measured. In molecular biology, we may restrict our consideration of "order" and "random" to that which appears in the study of sequences. Overman (1997) has written a very clear discussion of "order," "randomness," and "self-organization" for the general reader.

11.1.1 Do "orderly" sequences play a major role in genetics?

Let us consider again the two-headed coin problem I discussed in Chapter 8. The Guru has selected one sequence from a total of $2^{100} = 1.2676506002 \times 10^{30}$, namely, the sequence of 100 heads. This "orderly" and "nonrandom" sequence can be described by the algorithm in the computer: PRINT "heads" 100 times. We can select other sequences in the ensemble of 100: PRINT "heads-tails" 50 times. We may select more complicated sequences each requiring a longer description. However, there is such lack of "order" in this enormous ensemble of sequences that we must consider each sequence by itself.

The probabilities in problems of practical interest are almost always not all equal. Yet, the sequences generated by random walk or Markov processes (see the Mathematical Appendix) have a hierarchy of randomness that requires a measure. For this reason, the current definition of randomness must be replaced by one more sophisticated to come abreast of developments in mathematics. Consider the following two sequences in the binary alphabet [0, 1], (Chaitin, 1975):

010101010101010101010101

011011001101111 10001000

The first sequence clearly shows a pattern of orderliness and one is inclined to expect that the same pattern will persist indefinitely.

A pattern is not so clear in the second sequence. However, most gamblers have a system that they believe helps them beat the odds even though it can be shown that no system will do so (Feller, 1968). A gambler, looking for a system in the second sequence, will notice a clustering of 11 separated by one or two 0s. If he applies this system after the thirteenth event he will begin to lose. To save his system he will complicate and embellish it, but eventually he will lose. Fortuna will punish him for his impertinence by playing the Lorelei and will lead him to the gambler's ruin. By contrast, the first sequence, as well as the second, has an equal number of 0s and 1s and therefore they each have exactly the same *a priori* probability.

Here are several more sophisticated examples to show why we need a more careful means of distinguishing between "orderliness" and "randomness." Rational fractions are represented by an infinite sequence of decimal digits. The computer program for calculating the decimal sequences of rational fractions is very simple and short, for example, divide 17 by 39. The answer is the repeating sequence: 0.435, 897, 435, 897, 435, 897, 435, 897 . . . that, although it is an infinite sequence, clearly has an orderly pattern from which the rest of the sequence can be predicted. We can use the orderliness of this sequence to compose a message specifying an algorithm that generates the infinite sequence. "PRINT 0.435897, and repeat 435897 indefinitely." Thus a finite sequence can contain all the information in an infinite one. Some authors compare an organism to crystallization, especially snowflakes. We see from this that the information in crystals is: *repeat this pattern indefinitely*. Therefore, crystals contain very little information and have nothing to do with the origin of life.

During the preparation of tables of random numbers, the question arose whether the sequence of decimal digits representing transcendental numbers such as π and e could pass the standard tests for randomness. It was found that the decimal sequences representing some transcendental numbers did, indeed, pass such tests, π and e included (Parthia, 1962; Pincus and Kalman, 1997; Pincus and Singer, 1998; Stoneham, 1965). These sequences therefore exhibit no orderly pattern from which the rest of the sequence can be predicted. Yet, clearly, Urania, the Muse of Science did not select the numerical sequences representing π and e by a stochastic process. Because each digit is computable by an algorithm of finite information content, we may conclude, in general, that a finite message, specified by an algorithm may exist that carries all the information in an infinite sequence even though there is no discernible pattern.

One might be led to believe that, having a definition and a measure of randomness, one could prove a given sequence to be random. In fact, it is impossible to do so (Chaitin, 2001a). A sequence is not random if one finds a program that compresses the sequence to one substantially shorter than the sequence itself. We need not prove that the program is the shortest one, just that it is shorter than the original sequence. By contrast, to prove that a given sequence is indeed random one must prove that no shorter program exists.

Every computable number must be capable of being described by a finite number of symbols, that is, by an algorithm as in computer technology. Because a description or algorithm consists of a finite number of symbols there are only countable many algorithms. The total number of functions, say transcendental numbers, however, is transfinite, that is, uncountably infinite. There is, therefore, a transfinite number of functions or transcendental numbers that cannot be calculated by an algorithm (Hermes and Markwald, 1974). How can these considerations lead to a definition of randomness that does more than classify a sequence as random or not random and provides a measure of the amount of randomness in the sequence? An answer to this question emerged from the initially independent work of Solomonoff (1964), Kolmogorov (1958, 1965), Chaitin (1966, 1975, 1979), and Martin-Löf (1966) on the foundations of information theory and the foundations of probability theory. The current state of the theory is found in papers by Chaitin (1979, 1987a, 1987b, 1990), in Li & Vitányi (1997), and in Wolfram (2002).

The size in bits of the shortest computer algorithm describing the sequence or bit string is a measure of its Shannon entropy or randomness. The amount of randomness in π and e is just that in the shortest algorithm by which they are calculated. By the same token, we now have a definition and a measure of "order," namely, that an "orderly" sequence is one that can be specified by a short algorithm. For that reason, it is irresponsible for a scientist, even in a novel, to suggest that there is a message in the remote digits of π as Carl Sagan has done (Sagan, 1985).

One reason that "orderly sequences" can't play a major role in genetics is that there are so few of them. Let us estimate the number of sequences that have a large amount of randomness. Suppose we have compressed a sequence expressed in a binary alphabet (0, 1) suitable for the input of a computer. Let the length of the sequence be l. A very orderly sequence can be generated by a program of one symbol, either 0 or 1. There are two such programs. There are 2^2 programs that can be specified by two symbols, namely [(0, 0), (0, 1), (1, 0), (1, 1)]. In general there are 2^i programs specified by i symbols. The total number, S, of such programs is the sum of a geometric series:

$$S = 2^1 + 2^2 + 2^3 + \cdots 2^l$$
$$S = \frac{2(2^l - 1)}{(2 - 1)} = 2^{l+1} - 2.$$

The sum of the number of programs 10 less than l is $(2^{l-9} - 2)$. Thus for long sequences, there are a *factor* of 1,024 fewer programs in sequences just 10 shorter than l. Accordingly, almost all long sequences are indistinguishable from random ones (Wolfram, 2002). This fact may be unpleasant to some people. Gregory Chaitin (1999, 2001a, 2001b) has proved that *no procedure exists* to determine whether a given sequence can be calculated from a computer program. Consequently, it is impossible to determine whether a given sequence is random or not.

Schuster (2000) and Morowitz et al. (2000) object to "combinatorial explosion" leading to long sequences. They believe that self-organization, as a universal principle, will introduce "order" into diverse manifolds. "Orderly" sequences can not play a major role in genetics because there are so few of them in comparison with those that are random. Furthermore, because they can be described by a short algorithm, they cannot carry the information needed in most genomes (Yockey, 1981).

11.1.2 *Randomness and complexity*

Complexity is a concept that is widely misunderstood (Adami et al., 2000; Adami and Cerf, 2000; Behe, 1996) and widely viewed as being difficult to define. In other parts of the olive groves of academe, Kolmogorov (1958, 1965) and Chaitin (1966, 1979, 1999) have called *complexity* the Shannon entropy of the shortest algorithm needed to compute the sequence. Thus, both *random sequences* and *highly organized* ones are *complex* because a long algorithm is needed to describe each one (Wolfram, 2002). Information theory shows that it is *fundamentally undecidable* whether a given sequence has been generated by a stochastic process or by a *highly organized* one. This is in contrast with the classical law of the excluded middle (*tertium non datur*), the doctrine that a statement or theorem must be either true or false. Algorithmic information theory shows that truth or validity may also be *indeterminate* or *fundamentally undecidable*.

Gregory Chaitin (1990, 1999, 1992b, 2001a, 2001b) has shown that randomness is as fundamental in pure mathematics, as it is in the tossing of dice. Mathematical reasoning is the most powerful available to man yet we must leave to *Fortuna*, the goddess of chance, the fundamental limits of our knowledge as were the ancient Greeks and Romans.

> Chaitin (1979) discussed the role of complexity in biology: In discussions of the nature of life, the terms "complexity," "organism" and "information content" are sometimes used in ways remarkably analogous to the approach of algorithmic information theory, a mathematical discipline that studies the amount of information needed for computations. We submit that this is not a coincidence and that it is useful in discussions of the nature of life to be able to refer to analogous precisely defined concepts whose properties can be rigorously studied. We propose and discuss a measure of degree of organization and structure of geometrical patterns, which is based on the algorithmic version of Shannon's concept of mutual information. This paper is intended as a contribution to von Neumann's program of formulating mathematically the fundamental concepts of biology in a very general setting i.e. in highly simplified model universes.

The question of whether "complexity" increases along a phylogenetic chain can only be addressed when the well-established definition and quantitative measure of "complexity" such as that given by Chaitin and Kolmogorov is adopted in molecular biology (Yockey, 1977a, 1977b, 1977c, 1981, 1990, 1992, 2000, 2002). A DNA sequence of an alphabet of four letters *A, C, G, T* looks like a very long computer program (Ming Li and Vitányi, 1997). Another way of understanding information content as a measure of complexity is to say it measures the randomness of a sequence. Complexity is a scale with orderliness at one end and randomness at the other.

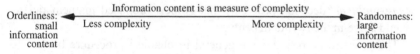

Figure 11.1. Relation of orderliness, randomness, and complexity.

The orderliness, so often misnamed "self-organization," is at the other end of complexity. There can not be an increase in complexity by "self-organization" as some authors have proposed (Füchlin and McCaskill, 2001; Lehn, 2002a. 2002b). When we speak of the amount of complexity in a sequence, we are speaking about the amount of its randomness as well.

The units of measurement of information content, orderliness, complexity, and randomness are the *bit* and the *byte*, which are familiar to computer users as a gauge of how much information their computers can process or store.

11.2 Algorithmic information theory and the isomorphism between communication systems, computers, the mathematical logic system, and the genetic logic system

Computer scientists (Landauer, 2000) have pointed out the relation or correspondence between the genetic logic system, computer programs, and communication systems such as the telephone and telegraph (see Section 2.1.1, Section 3.3, Section 5.1.1, and Figure 5.1). In Section 2.2.1, I discussed the criteria for *isomorphism* between systems (Ornstein, 1970). Examples of isomorphism are ubiquitous in biology. The science of taxonomy is based on the isomorphism of organs. The wings of a bat, the wings of a bird, the paws of a gorilla, and the hands of Vladimir Horowitz are isomorphic although their function is quite disparate. Let us now use this more general and exact terminology for relation or correspondence and point out in addition the *isomorphism* between the genetic logic system, the logic systems of communication systems, the logic systems of computers, and the logic system in mathematics by which theorems are proved from a list of axioms. These systems may be regarded as being the same abstract system (Ornstein, 1989). This is easily understood by the computer-literate. Now that includes almost everyone.

The basic principle on which computers operate is that of the Turing machine (Turing, 1936). Alan Turing (1912–54) conceived this abstract model of a computing machine to solve problems in the foundations of mathematics. In particular, he addressed the question of computable numbers and the application to David Hilbert's tenth problem, known as the *Entscheidungsproblem* (decision problem) that Hilbert proposed mathematicians would solve in the twentieth century.

The problem is to find some general mechanical procedure by which, *in principle*, all problems of mathematics formulated in axiomatic systems could be solved. In addressing the problem, Turing imagined an abstract machine in which a message or sequence is recorded on an input tape that could be weightless and of infinite length. In computer terminology, these messages or sequences are called *bit strings*, because they are expressed in a string of the computer alphabet (0,1). The algorithmic information content of these bit strings is, of course, measured in bits. There is a reading head that may move in either direction along the tape to read the input, which interacts with a finite number of internal states. These states are called a program in modern computer terminology. The program carries out its instructions on the message read from the input tape. The results are read out on an output tape and the machine stops when the program is executed.

The problem of deciding whether an algorithm will halt, thus completing its calculation, on a universal Turing calculating machine is related to Gödel's Incompleteness Theorem (Gödel, 1931, 1992). That is, the question of whether the machine will eventually solve the problem or will go on indefinitely was proved by Turing (1936) to be unsolvable in general. If the Turing machine stops, the number is computable, otherwise the number may be random and incomputable. By the same token, no computable number is random. A more detailed description of Turing machines and their use in the study of the computability of numbers would be diversion. A rather detailed and accurate account of Turing machines is given in the popular book by Penrose (1989).

The logic of Turing machines has an isomorphism with the logic of the genetic information system. The input tape is DNA and the bit string recorded is the genetic message. The internal states are the tRNA, mRNA, synthetases and other factors that implement the genetic code and constitute the genetic logic system. The output tape is the family of proteins specified by the genetic message recorded in DNA. There also is an isomorphism between the information in the instructions on the Turing machine tape and the information in the list of axioms from which theorems are proved.

Without noticing these isomorphisms, corresponding properties would appear to be unrelated. But in each of these four cases one has an information source, a transmission of information, a set of instructions or tasks to be completed, and an output. By the same token, we shall find examples of properties in each system to have corresponding properties in the others. I have taken advantage of these isomorphisms in communication systems and computers and apply them to problems in molecular biology.

11.2 The unknowable and the impossible

11.2.1 The unknowable in mathematics

Are there limits in our ability to reason from axioms as there are limits of accuracy in measuring, weighing and counting? David Hilbert, the famous mathematician, delivered the last lecture of his career: *Logic and the understanding of nature* on September 8, 1930. He ended his lecture with these words: *"Wir müssen wissen! Wir werden wissen!* [We must know We shall know]. He was soon to be disappointed. Mathematics is the most capable means of human reasoning, so it may come as a surprise that some things are *impossible, unknowable,* and *undecidable* even in mathematics (Chaitin, 1990, 1992a, 1992b, 1999, 2001a, 2001b; Overman, 1997; Overman and Yockey, 2001).

Computer programmers are aware that no procedure exists to determine whether a given computer algorithm will halt, that is, complete its calculation (Chaitin, 1979). Nevertheless, the halting property exists whether it is *knowable* to us or not. Alan Turing, a mathematician and cryptographer, discovered this before any computers had been built (Turing, 1936).

The ancient Greek mathematicians virtually completed plane and solid geometry. Euclid's *Elements* was used as a textbook until the late nineteenth century. There are three classical problems they could not solve, (Bold, 1982):

The trisection of a given acute angle;

The doubling of the cube;

The construction of a square of area equal to that of a given circle. The only instruments allowed are the straight edge and the compass.

It is easy to bisect any given angle. At first sight it appears that there also must be a way to trisect a given angle, because after all, it is simple to construct an angle that is triple that of any given one. We can measure the given angle and construct one a third in size. But that is an act of measurement subject to error, not one of pure reason.

There are explicit solutions in terms of the coefficients for general equations up to the fourth degree but none for the fifth and higher. Urania, the Muse of Science, has neglected to provide mankind a means to solve them. But if one substitutes numbers for each coefficient the solutions pop in place and may be found to as great an accuracy as one wishes. They exist but are hidden from us. Thus, there is a distinction between *unknowable* and *impossible.*

There are only five Platonic regular solids; it is *impossible* to make any others:

1. the tetrahedron, four sides, the faces of which are triangles,
2. the hexahedron or the cube, six sides, the faces of which are squares,
3. the octahedron, eight sides. the faces of which are triangles,

4. the dodecahedron, twelve sides, the faces of which are pentagons,
5. the icosahedron, with twenty sides, the faces of which are triangles.

Sir Isaac Newton found that he could calculate exactly the motion of one planet about a central Sun attracted by the force of gravity between them. When a third body is introduced, we find that no procedure exists to solve the three-body problem (Wolfram, 2002). Nevertheless, the planets go about their orbits unconcerned with the limitations of mankind's mathematics.

Einstein (Pais, 1982) could never give up strict Aristotelian causality or accept the stochastic character of quantum mechanics. He did not believe that God casts lots with the world. Perhaps so, nevertheless, we can calculate by means of quantum mechanics only *probability amplitudes* and this is a fundamental limit of our knowledge of atomic systems. Thus, quantum transitions are *unknowable* to us but they are not *impossible*. They occur exponentially in time and that proves their appearance is random.

Some people may believe that quantum mechanics is too esoteric to have any relation to ordinary science and to our affairs. On the contrary, the laser that records items we purchase at the supermarket is a totally quantum mechanical device. Einstein found its operation in the mathematics of quantum mechanics. There was a long wait before its discovery in the laboratory. There is no path by reasoning from classical spectroscopy to quantum mechanics and the laser.

11.2.2 The unknowable origin of the genetic code

In Chapter 7 we found that the genetic code is much like all codes used in communication. Looking backward in time, through a glass darkly, how near to the origin of the genetic code can we see? It is often speculated that the genetic code began as a binary alphabet. This is not a fruitful speculation as DNA and RNA are composed of four compounds that form a primary four letter alphabet. Even with some imagination we may consider a sixteen-member code composed of the doublets of UCAG. The origin of the genetic code is unknowable. I have no doubt that if the historic process leading to the origin of life were *knowable* it would be a process of physics and chemistry. Thus, the process of the origin of life is *possible* but *unknowable*.

11.2.3 The unknowable in evolution and horizontal gene transfer

Looking back in time, we must consider the simplest organism that could have survived, perhaps in the early ocean. We can get some guidance from those organisms with the smallest genomes that survive today. Clearly, there must be some essential "housekeeping" tasks that involve the biochemistry that forms the minimum number of amino acids, nucleotides, and other essential molecules (Gil et al., 2002; Sassetti, Boyd, and Rubin, 2001). The *Mycoplasma (class Mollicutes)* are considered the representatives of a minimal cell (Mushegaian and Koonin, 1996). *Mycoplasma genetatilium*, with 517 genes, is the smallest autonomously replicating organism that has been completely sequenced (Hutchinson et al., 1999). Several proteins in that organism are conserved to man (Balasubramanian et al., 2000).

Horizontal gene transfer between organisms was pervasive and dominating in the early history of life. About forty genes are found to be exclusively shared by humans and bacteria. They are examples of horizontal gene transfer from bacteria to vertebrates (Salzberg, White, Peterson, and Eisen 2001). Gene transfer occurs between a mitochondrion and the nucleus. For that reason, many genes give believably different phylogenies for the same organism (Doolittle, 1999). The nearly universal structure of the genetic code and the handedness of proteins and nucleic acids is preserved in horizontal gene transfer and attests to a universal ancestor. Nevertheless, horizontal gene transfer has made a jumbled network more like a thicket than a tree and has substantially erased the record of the earliest genetic sequences (Archibald et al., 2003; Bushman, 2002; Kinsella et al., 2003; Kurland et al., 2003; Lake, Jain, and Rivers, 1999; Syvanen and Kado, 1998; Woese, 2002). Its vagaries limit the ability of genomic sequencing to follow the phylogenetic tree to the progenote or universal ancestor (Bushman, 2002; Doolittle, 2000; Lawrence and Ochman 1998; Woese, 1998; Woese et al., 2000). The universal phylogenetic tree exhibits the relationship of all organisms, those from which extant organisms evolved and those to be evolved in the future (Woese, 1998, 2000, 2002). The root of the phylogenetic tree is actually a Darwinian Threshold that represents the first stage in molecular evolution. As one attempts to follow the tree to its Darwinian Threshold by vertically derived sequences one encounters the effects of horizontal gene transfer as the principal driving force in early cellular evolution (Woese, 2002). This means that the earliest branches of the tree or net are not knowable.

11.2.3 The relation of "order" and "complexity" to the origin of life

Mathematicians (Hilbert, 1902) who thought that all mathematical statements could be either proven or disproved were astonished by the "incompleteness" theorem of Kurt Gödel that for any axiom system that is consistent and can be expressed in a computer program there are statements that can be neither proved nor disproved (Gödel, 1931; in English translation, 1992). That is, they are *undecidable* to mathematicians. In Section 1.1.3, I reminded the reader that Niels Bohr (1933) proposed that life is *consistent* with but *undecidable* or *unknowable* by human reasoning from physics and chemistry. I showed in Chapter 8 that Darwin did not believe that life emerged in a "warm little pond." Darwin believed that the origin of life is *unknowable* or *undecidable*.

There is no requirement that Nature's laws be plausible or even known to mankind. As Hamlet said to his friend: *"There are more things in heaven and earth, Horatio, than are dreamt of in your philosophy"* (Hamlet, act 1, scene V). John Burdon Sanderson Haldane (1892–1964) wrote in 1927: *"The universe is not only queerer than we suppose, but queerer than we can suppose."* Thus, although some are optimistic that life may be made in the laboratory (Deamer 1997), it may well be that scientists, by *counting, measuring, or weighing and employing the calculating or reasoning element in the soul* (Plato, 428–348 B.C., *The Republic*, Book X) will come closer and closer to the riddle of how life emerged on Earth but, because of the limitations of measuring and human reasoning, like Zeno's Achilles, will never achieve a complete solution (Zeno of Elea, c. 490 B.C.).

12

Does evolution need an intelligent designer?

... when we come to inspect the watch, we perceive ... that its several parts are framed and put together for a purpose, e.g. that they are so formed and adjusted to produce motion, and that motion so regulated as to point out the hour of the day; that if the different parts had been differently shaped from what they are, or placed after any another manner or in any other order than that in which they are placed, either no motion at all would have been carried on in the machine, or none which would have answered the use that is now served by it ... the inference we think is inevitable, that the watch must have had a maker – that there must have existed, at some time and at some place or other, an artificer or artificers who formed it for the purpose which we find it actually to answer, who comprehended its construction and designed its use.

William Paley (1743–1805)

12.1 Does the complexity of biology call for intelligent design?

Creationists believe that the enormous complexity of biology, and of the universe in general, calls for an Intelligent Designer (Behe, 1996; Behe, Dembski, and Meyer, 2002; Dembski, 1998a, 1998b, 1999, 2002; Dembski and Ruse, 2004).

St. Thomas Aquinas (1225–74), in his *Five Proofs*, referred to the appearance of design in the universe as the fifth proof of the existence of God. David Hume (1711–76) is also known for his discussion of the same argument, published three years after his death. Hume wrote in the form of a dialogue between three characters. Philo, Cleanthes, and Demea. The design argument, as a sufficient foundation for religion (but not necessarily the belief of Hume), is spoken by Cleanthes in Part II of the Dialogues:

> The curious adapting of means to ends, throughout all nature, resembles exactly, though much exceeds, the production of human contrivance, or human design, the thought, wisdom and intelligence. Since therefore the effects resemble each other, we are led to infer, by all the rules of analogy, that the causes also resemble; and that the Author of nature is somewhat similar to the mind of man; though possesses of much large faculties, proportioned to the grandeur of the work executed.

176

Today the most well-known comment on Intelligent Design evolved (please excuse!) from the writings of the Reverend William Paley (1743–1805), a priest of the Church of England. His publication, twenty-two years after Hume, discussed finding a watch upon crossing a heath (Paley, 1802). The watch, of course, must have been made by a watchmaker for a purpose. Following the rules of analogy, Paley pursued that point with regard to the eye. How much more complex is the eye, and all living things, than a watch!

The evolution of vision concerned Darwin (1872, Ch. VI):

Organs of Extreme Perfection and Complication

> To suppose that the eye with all its inimitable contrivances for adjusting the focus to different distances, for admitting different amounts of light and for correcting spherical and chromatic aberration, could have been formed by natural selection, seem, I freely confess, absurd in highest degree.

Eyes based on the camera principle have appeared, almost universally in vertebrates, but also in cephalopods and in annelids. Vision also occurs in insects, spiders, lobsters, and so forth, but by the compound eye, a different optical principle (Mayr, 1982).

12.1.1 Darwin and the clergy

"Darwinism" was and still is believed by some members of the clergy to be heresy that challenges Christian orthodoxy and the veracity of the creation story in Genesis. Both the Roman Catholic Church and the Church of England regard heresy, apostasy, and atheism as very serious matters to be opposed in every way possible. The justification is Matthew 7:15: *Beware of false prophets who come to you in sheep's clothing but inwardly are ravenous wolves.*

The British Association for the Advancement of Science met at Oxford University, in June 20, 1860, to discuss Darwin's book, *The Origin of Species by Means of Natural Selection or the Preservation of Favored Races in the Struggle for Life*. Darwin, who had little appetite for involvement in the growing controversy, was chronically ill. Thomas Henry Huxley (1825–95), known as "Darwin's Bulldog," who relished the intense debate, defended Darwin's view of evolution. Samuel Wilberforce (1805–73), a bishop of the Church of England, made a systematic assault on evolution. Warming to his subject, he turned to Huxley and asked if it was his grandfather or his grandmother who was descended from an ape. Huxley muttered to himself, "The Lord hath delivered him unto mine hands." Huxley replied that he would rather be descended from a monkey than from someone who would so prostitute the truth.

From the days of William Jennings Bryan (1860–1925) and the Scopes trial, in Dayton, Tennessee, July 1925, the antievolutionists, biblical creationists, and creation-science advocates have attempted to prevent biology teachers from teaching or discussing "Darwinism" in the public schools. They regard evolution as "just a theory," implying that it is merely a speculation, and has serious scientific difficulties. They advocate "balanced treatment for creation science and evolution." The Reverend Dr. Jerry Falwell, executive pastor of the Thomas Road Baptist Church and a leader of the Religious Right, is particularly active against teaching of evolution. One may presume that evolution is not taught at Liberty University. Philip E. Johnson, a professor of law at the University of California at Berkeley (Johnson, 2000), is a leader in the Intelligent Design movement. By contrast, many of the clergy believe that "The Bible teaches us how to go to Heaven, not how the heavens go."

12.1.2 Michael J. Behe and the "irreducible complexity" of life

Michael J. Behe, an associate professor of biochemistry at Lehigh University, has received considerable attention from his book *Darwin's Black Box* (Behe, 1996). Fifty years after the discovery of the function of DNA, Behe regards the function of the genetic message as so much biochemistry. His objection to evolution is that biology is "irreducibly complex."

> By *irreducibly complex* I mean a single system composed of several well matched, interacting parts that contribute to the basic function, where in the removal of anyone of the parts causes the system to cease functioning. . . . And as the number of unexplained irreducibly complex biological systems increases, our confidence that Darwin's criterion of failure has been met skyrockets toward the maximum that science allows.

His example of "irreducible complexity" is a mousetrap that must have all its parts functioning and therefore, like Paley's watch, must have been formed by Intelligent Design.

Neither Paley's watch nor Behe's mousetrap is alive. They do not heal themselves nor do they produce little watches or little mousetraps. I replied

(Yockey, 2001) that the Venus flytrap (*Dionaea muscipla*) and other carnivorous plants are alive, they reproduce, and are products of evolution (Merbach et al., 2002; Surridge, 2002). Their genomes are recorded in the same DNA as are all other organisms. According to the analogy to a Turing machine, the protein sequences that are read out from DNA are isomorphic to a computable number. If the Turing machine stops, the number is computable, otherwise the number may be random and incomputable. By the same token, no computable number is random. Although it is impossible to determine if a *given* sequence is random or not, the protein sequences that compose living organisms are not random or "irreducibly complex."

Carnivorous plants dispose of the victim, a feature that is lacking in Behe's mousetrap. (One would suppose that an associate professor of biochemistry would have thought of carnivorous plants.)

12.2 Charles Darwin and Gregor Mendel

The laws governing inheritance are for the most part unknown. No one can say why the same peculiarity in different individuals of the same species, or in different species, is some times inherited and sometimes not so; why the child often reverts in certain characteristics to its grandfather or grandmother or more remote ancestor; why a peculiarity is often transmitted from one sex or to both sexes, or to one sex alone, more commonly to but not exclusively to the like sex. (Darwin, 1872)

Darwin, in this passage, expressed the essential facts, but he did not know the laws of genetics because he had not read Gregor Mendel's *Versuche über Pflanzen-Hybriden* (Experiments on Plant Hybrids) (1865). Mendel (1822–84) visited the Great Exhibition in London in 1862. Mendel did not know of Darwin's work at that time. A marked copy of the German edition of *The Origin of Species* was found among his books upon his death. Darwin's movements are known precisely from his records. These two contemporary geniuses were never in the same town at the same time.

Darwin believed that characteristics are blended (Fisher, 1930). His analysis depended entirely on morphology. The comparative morphology of chimpanzees, gorilla, orangutan, and Old World monkeys have intrigued as well as annoyed us. Now that DNA is being sequenced for all organisms from viruses to man, the relation of all organisms can be based strictly on the amount of similarity, that is mutual information, between the DNA sequences.

It is now possible to make a comparison, in numerical form, between all these animals and man that is beyond question. The sequences of cytochrome c in the chimpanzee and human are identical, not only in the proteins but, moreover, in the codons themselves. This heirloom in our genetic attic has come to us in identical form as a message from our common ancestor, whether we like it or not (Clark et al., 2003; Penny, 2004).

Furthermore, Wildman et al. (2003) have shown in a rather detailed study that the two chimpanzee species, *Pan troglodytes* (common chimpanzee), and *Pan paniscus* (bonobo chimpanzee), have DNA sequences that are 99.4 percent identical with *Homo sapiens*. They propose that all three of these species be included in genus *Homo*, and of course *Homo neanderthalensis*, the Neanderthal man (Harvati, Frost, and McMulty, 2004), a separate species for five hundred thousand years (Ponce de León Marcia and Zollikofer, 2001).

12.3 Evolution and error in the genetic message

12.3.1 Error processes in the genome

Natural processes conspire to accumulate errors in the genetic message. The Arrhenius equation describes how molecules may pop from one energetically stable state to another by classical thermo-agitation. Let t be the time of the transition, W, the energy barrier, τ the time constant, T the absolute temperature, and k the Boltzmann constant

$$t = \tau e^{W/kT}.$$

The quantum mechanical tunneling through an energy barrier that George Gamow showed to be responsible for alpha radioactivity also contributes to the molecular transitions that create error in the genetic message (McMahon, 2003; Zuev et al., 2003). This process causes amino acids and nucleic acids to become racemic over the terrestrial geologic time scale of five to ten million years.

Chemical reactions due to the tunneling process arise from wave-particle duality, a peculiarity of quantum mechanics. The quantum mechanical wavelength of the electron is large, due to its small mass, so that electron tunneling is manifested in chemical reactions. The much larger mass of the proton means a much smaller quantum mechanical wavelength. Nevertheless, quantum mechanical tunneling is responsible for some chemical reactions and loss of information due to errors in the genetic message over evolutionary times.

12.3.2 Majority logic redundance and redundance in the genetic code

I have argued that the origin of life, like the origin of the universe, is unknowable. But once life has appeared, Shannon's Channel Capacity Theorem (Section 5.3) assures us that genetic messages will not fade away and can indeed survive for 3.85 billion years without assistance from an Intelligent Designer. As I pointed out in Section 5.1.4, there is an enormous redundance in protein families. Although majority logic redundance plays an unimportant role in telecommunications it is controlling in the genetic communication system. This shows without a doubt that evolution and genetics cannot be understood except by information theory.

Evolution of the genetic code and of the genome proceeds by a Markov chain or random walk (see the Mathematical Appendix). There are a number of events in this Markov chain including gene duplication and horizontal gene transfer (Section 5.1.4) through biosynthetic pathways (Chapter 7). The duplicated genes provide redundance protection against mutation to a nonspecific message, but also to the evolution of a new gene (Graure and Li, 2000; Haldane 1932). The slight modifications, as Darwin believed, are supplemented by gene duplication and horizontal gene transfer. Current results in genetic sequencing show that duplicated genes are abundant in most genomes (Lynch 2002; Lynch and Conery, 2000).

The messages in the DNA sequences are similar to programs in modern computer terminology. mRNA acts like the reading head on a Turing machine that moves along the DNA sequence to read off the genetic message to the proteome. The fact that the sequence has been read shows that it is not "irreducibly complex" or random. By the same token, Behe's mouse trap is not "irreducibly complex" or random.

The same genetic code, the same DNA, the same amino acids and the genetic message that unites all organisms, independent of morphology, proves that the theory of evolution is as well established as any in science. There is indeed Aristotle's "Great Chain of Being" (Lovejoy, 1936) that relates all living things. How this happened must be learned by measuring, counting and weighing as Socrates taught us.

The fact that there are many things unavailable to human knowledge and reasoning, even in mathematics, does not mean that there must be an Intelligent Designer.

13

Epilogue

Although I am fully convinced of the truth of the views given in this volume under the form of an abstract, I by no means expect to convince experienced naturalists whose minds are stocked with a multitude of facts all viewed, during a long course of years, from a point of view directly opposite to mine. It is so easy to hide our ignorance under such expressions as the "plan of creation," "unity of design," &c., and to think that we have given an explanation when we only re-state a fact. Any one whose disposition leads him to attach more weight to unexplained difficulties than to the explanation of a certain number of facts will certainly reject the theory. A few naturalists, endowed with much flexibility of mind, and who have already begun to doubt the immutability of species, may be influenced by this volume; but I look with confidence to the future, – to young and rising naturalists who will be to able view both sides of the question with impartiality.

Charles Darwin (1872, p. 444)

13.1 The Central Dogma and the origin of life "proteins first"

One of the more important contributions in this book, and in two previous papers (Yockey, 2002a, 2002c) is to point out that no code exists to send information from protein sequences to sequences in mRNA or DNA. Therefore, it is *impossible* that the origin of life was "proteins first" from Haeckel's *Urschleim* (Section 3.1.1). The restrictions of the Central Dogma on the origin of life are mathematical (Battail, 2001; Yockey, 1974, 1978, 1992, 2000, 2002a). Scientists cannot get around it by clever chemistry. This restriction prevails in spite of the concentration of protein in a "prebiotic soup" may have been or may be on some "Earth-like" planet elsewhere in the universe. For this and other reasons the origin of life is unknowable.

Most of the publications on the origin of life are based on chemical reactions that under "plausible prebiotic conditions" (as though mankind is capable of deciding what is plausible") are presumed to provide an *Urschleim* from which proteins are alleged to have originated in the early ocean. Ernst H.P.A. Haeckel (1834–1919) published that speculation in 1866. It is usually ascribed to Oparin who was not born until 1894.

NASA is totally committed to the Oparin dialectical materialist speculations on the origin of life. Most of the origin of life projects supported by

NASA and other funding agencies are "proteins first" and are due to go the way of perpetual machines. They may produce interesting chemistry but they have nothing to do with the origin of life.

By contrast, the so-called *reverse transcription* that astonished many people is in accordance with the discussion of the Central Dogma and the mathematical theory of redundant codes (Yockey, 1974, 1978, 1992, 2000, 2002). There would have been very much less *Sturm und Drang* in the olive groves of academe if that knowledge had been applied to understanding the genetic code.

It is a characteristic of the True Believer in religion, philosophy, and ideology that he must have a set of beliefs, come what may (Hoffer, 1951). Belief in a primeval soup on the ground that no other paradigm is available is an example of the logical *fallacy of the false alternative*. In science, it is a virtue to acknowledge ignorance. There is no reason that this should be different in the research on the origin of life. The best advice that one could have given to the alchemist would have been to study nuclear physics and astrophysics, although that would not have been helpful at the time. We do not see the origin of life clearly, but through a glass darkly. Perhaps the best advice to those who are interested in the origin of life would be to study biology as Simpson (1964) proposed.

13.2 New axioms in molecular biology and genetics

The sequence hypothesis in the genome and in the proteome is a new axiom in molecular biology. Mayr (1982, 1988), a philosopher of biology, states that it is unique to biology for there is no trace of a sequence determining the structure of a chemical or of a code between such sequences in the physical and chemical world. The sequence hypothesis and the genetic code are in addition to but do not transcend the laws of physics and chemistry. This is quite different from the nineteenth-century doctrines of vitalism and dualism whose proponents believed that there are processes unique to living organisms that are *contrary* to the laws of physics and chemistry (Mayr, 1982; Pauly, 1987).

Galileo (1554–1642) believed that the language of Nature is inherently mathematical and is essential to describing natural phenomena. Although there are many fields of biology that are essentially descriptive, with the application of information theory, theoretical biology can now take its place with theoretical physics without apology. Thus, biology has become a quantitative and computational science as George Gamow suggested. By employing information theory, comparisons between the genetics of organisms can now be made quantitatively with the same accuracy that is typical of astronomy, physics, and chemistry.

13.3 Eigen's "error catastrophe," Intelligent Design, and the survival of life for 3.85 billion years

Darwin's theory of evolution, often called "Darwinism," in its modern form, requires the transmission of the essential parts of the message in the genome of Pooh-Bah's ancestor, a *protoplasmal primordial atomic globule*, for more than 3.85 billion years since the origin of life (Mojzsis et al., 1999; Woese, 2000; Yockey, 1977b, 2002). That could not happen if noise obscured the genetic message, as it would according to Eigen's "error catastrophe" (Section 10.3).

Shannon's Channel Capacity Theorem (Shannon, 1948), has firmly established the conditions by which messages may be communicated with minimum error. I have argued that the origin of life, like the origin of the universe, is unknowable. But once life has appeared, Shannon's Channel Capacity Theorem (Section 5.3) assures us that genetic messages will not fade away, like old soldiers, and can indeed proliferate and survive for 3.85 billion years without assistance from an Intelligent Designer (Mojzsis et al., 1999; Woese, 2000; Yockey, 1977b, 2002).

As I pointed out in Section 5.1.4, there is an enormous redundance in protein families. Although majority logic redundance plays an unimportant role in telecommunications, it is controlling in the genetic communication system. The genetic information system operates without regard for the specificity of the message because it must be capable of handling *all* genetic messages of *all* organisms, extinct and living, as well as those not yet evolved. That is possible *only* because the message in the genome is *segregated, linear, and digital*. This shows without a doubt that evolution and genetics can not be understood except by information theory.

There are no "irreducibly complex systems" in biology and there is no need for an Intelligent Design in evolution. The genetic system operates much like a computer, measuring the genetic message in bits and bytes. The protein sequences that are read out from DNA are isomorphic to a computable number, according to the analogy to a Turing machine. Therefore, they are not random or "irreducibly complex." Although it is impossible to determine if a *given* sequence is random or not, the protein sequences that compose living organisms are not random or "irreducibly complex." Their complexity is measured by the number of bits or bytes that describe the sequence. That puts "Darwinism" on a quantitative foundation as firm as any in the physical sciences.

13.4 Orderly sequences, the Second Law of Thermodynamics, and evolution

An accomplishment of this book is to give a quantitative measure to the terms "random," "nonrandom," "order," and "complexity" to discussions about biology. "Orderly" sequences in DNA do not play a role in genetics, only those that are complex do so. All attempts to find "order," where there is no "order," are futile. One reason that "orderly sequences" can't play a major role in genetics is that there are so few of them.

A number of authors confuse Shannon entropy of probability theory with Maxwell–Boltzmann–Gibbs entropy of statistical mechanics. Contrary to Schrödinger (1987, 1992), Wiener (1948), Eigen, (1992), and a number of authors – their names are Legion for they are many – Shannon entropy is not *negentropy*. Life does not feed on *negentropy* (Pauling, 1987) as a cat laps up cream.

An accomplishment of this monograph is to shoot down that *canard ancien*, now more than one hundred years old, which states that evolution is incompatible with the Second Law of Thermodynamics because evolution creates "order," whereas the Second Law demands that "disorder" increase with time in isolated systems. I have shown that this belief is merely a play on words. The correct explanation shows that an increase in Kolmogorov–Chaitin algorithmic entropy is *required* for evolution to proceed (Chaitin, 1979).

13.5 The Shannon–McMillan–Breiman Theorem

The correct way to calculate the number of sequences in a family of nucleic acid and polypeptide chains is by the Shannon–McMillan–Breiman Theorem: 2^{NH}. The expression $(20)^N$, usually found in the literature, contains an enormous number of "junk" sequences.

13.6 The Human Genome Project and sequencing the genomes of other organisms

The Human Genome Project, an international effort of thirteen years, released the first reference sequence of the human genome in April 2003. The genomes of a number of other organisms are also being published in the scientific journals. Now that DNA is being sequenced for many organisms from viruses to man, the relation of selected organisms no longer depends only on the interpretation of morphology but can be based strictly on the amount of similarity, that is mutual entropy, in the DNA and protein sequences.

13.7 The origin of life, religious apologetics, and the unknowable

> Surely one of the most marvelous feats of 20th-century science would be the firm proof that life exists on another planet. All the projected space flights and the high costs of such developments would be fully justified if they were able to establish the existence of life on either Mars or Venus. In that case, the thesis that life develops spontaneously when the conditions are favorable would be far more firmly established, and our whole view of the problem of the origin of life would be confirmed. Stanley L. Miller & Harold C. Urey, *Science* **130**, p. 251 (A. D. 1959), 1939, pp. 245–51

As George Gaylord Simpson (1964) pointed out, many authors who write about the origin of life do not realize that they are writing fiction, humor, or religious apologetics. The Fathers of the Church who established the Nicene Creed speaking *ex cathedra* knew they were establishing a religious creed. Miller and Urey, also speaking *ex cathedra*, asked Destiny for confirmation of their faith. The editors of the journal *Science* did not realize that they were publishing religious apologetics.

Space exploration has eliminated Venus from the search for extraterrestrial life, but NASA and the space exploration of other countries continues the search on Mars in spite of the decidedly negative results of the Viking Missions in the summer of 1976. While finding evidence of life on Mars would confirm the views of Ernst H.P.A. Haeckel (1834–1919), Miller (1920–), and Urey (1893–1981), by contrast, it is unlikely that True Believers, would accept evidence that life *never* existed anywhere on Mars at any time.

Convincing proof that biological material yet to be found on Mars, or originated on Mars, or at least was not from Earth must be that it be composed of *both* right-handed amino acids and left-handed ribose sugars of DNA and RNA. That is unlikely to happen because these compounds racemize in a time much shorter than the geological time scale of several million years. *The abscence of evidence is evidence of abscence.*

The history of the "life on Mars" effort that played a primary role in the Mariner and Viking space craft missions and the current ones shows an intemperate *will to believe* by leading scientists who saw in their data what they wished to see (Horowitz, 1986, 1990). The usual creative skepticism of scientists will not play a role until the enormous weight of evidence shows that belief that there is or was life on Mars is based on faith.

According to de Duve:

> Life is increasingly explained in terms of the laws of physics and chemistry. Its origin must be accounted in similar terms. The universe is awash with life.
>
> My conclusion after consideration of the underlying chemistry is that given the opportunity, the development of life is very likely to take the course it actually took, at least in essential aspects. (de Duve, 1995)

Let us remind ourselves that the laws of physics and chemistry are much like the rules of grammar. They must be obeyed, but there is not enough information in the rules of grammar to produce Lincoln's Gettysburg Address. The evolution of life from Pooh-Bah's *protoplasmal primordial atomic globule* is a Markov process or random walk leading to the chain of events we call evolution. The opinion of de Duve that evolution would take the same course, on some other planet elsewhere in the Universe, it took on Earth is based on faith. If life is a "cosmic imperative" and "the universe is awash with life" (de Duve, 1995), that is unknowable to us. The comments of de Duve and those of Miller and Urey must be regarded as religious apologetics. The prevalence of humanoids or little green men in the Milky Way Galaxy is unknowable (Simpson, 1964).

Contrary to current opinion it is not *deterministic* but, rather, it is *undecidable* that life would almost inevitably arise spontaneously from nonliving matter on young planets "sufficiently similar" to Earth elsewhere in the universe. Nevertheless, a search for extraterrestrial intelligence (SETI) has been conducted by the Phoenix Project lead by Dr. Frank Drake of the SETI Institute in Mountain View, California. Richard A. Kerr in *Science* (2004) quoted Dr. Drake that the Phoenix Project found nothing after listening to radio broadcasts from more than 700 Sun-like stars within 150 light-years of Earth.

Although it is now more than fifty years since the Watson and Crick discovery of the role of DNA, and the genetic code, there is nevertheless, considerable work attempting to find the transition from nonliving matter to living organisms (Szostak, Bartel, and Luisi, 2001). Two "workshops" were held to consider the "transition from nonliving to living matter" (Rasmussen et al., 2004), one at Los Alamos National Laboratory and another at Dortmund, Germany, the seventh European Conference on Artificial Life. Those who make such attempts will "sadden after none or bitter fruit."

This is not unusual in science. According to J. Robert Oppenheimer: "Classical mechanics, when applied to atoms, is not so much wrong as inappropriate:"

> The recognition of the essential importance of fundamentally atomistic features in the function of living organisms is by no means sufficient, however, for a comprehensive explanation of biological phenomena, before we can reach an understanding of life on the basis of physical experience. (Niels Bohr, 1933)

A great deal of effort has been expended in finding theories (i.e., algorithms) for the origin of life without success. The reason may not be that we are not smart enough or that we have not worked hard enough. The reason may be that no structure or pattern exists which can be put into the terms of an algorithm of finite complexity. The reader who has solved Socrates' problem (discussed in Plato's dialogue *Meno*) of doubling the square will find his or her efforts to double the cube futile. This does not mean that cubes twice the size of any given cube do not exist. It means that the *solution* to the problem is *undecidable*; it is beyond human reasoning. Gödel (1931) proved that there are theorems in number theory that are true but cannot be proved and there are propositions that are *undecidable* from the axioms of number theory.

Because creative skepticism is the cardinal virtue in science, one would expect that proponents of the primeval soup paradigm would be searching actively for direct geological evidence of such a condition in the early ocean. The power of ideology to interpose a fact-proof screen (Hoffer, 1951) is so great that this has not been done (perhaps for fear that its failure may be exposed). Bungenberg de Jong (1893–1977) wrote a long paper

Coacervation and its meaning for Biology) but his coacervates were gelatin, already biological.

Darwin did not believe in a "warm little pond" from which life is often alleged to have emerged by "chemical evolution" (Yockey, 1995, 2000, 2002a). Had he thought the "warm little pond" a worthy idea, he certainly would have published it.

Mathematical appendix

MA1.1 The origins and interpretation of probability

The origin of the primitive ideas of chance and uncertainty seems to be lost in history. Chance and uncertainty were poetically personified in the Greek pantheon by the goddess Tyche. The Romans borrowed her and renamed her Fortuna. The Fates, three goddesses who controlled human life, no doubt were more ancient. They were Clotho, who spins the thread of life (now known to be DNA); Lachesis, who casts lots to determine how long life shall be; and Atropos, who cuts the thread of life. These goddesses are sometimes referred to collectively and pejoratively as "blind chance" or "mere chance." They may be called vindictive, callous, cruel and fickle, but not blind (see gambler's ruin; Feller, 1968).

In our sophisticated modern world, we still cast lots to determine events, especially those in which we wish to avoid responsibility. One hears of "calculated risk" from politicians, military officers, and business executives when the strategy has gone awry and of shrewd judgment when things have gone well; thus, they take the credit for success while putting the responsibility for failures on the fickleness of Fortuna, Tyche or the three Fates, Clotho, Lachesis, and Atropos, as did the ancient Greeks and Romans. We often say "chance" when we mean "opportunity."

MA1.1.1 Do the Gods cast lots?

Einstein (Pais, 1982) could never give up strict Aristotelian causality or accept the stochastic character of quantum mechanics. Therefore, he did

191

not believe that God casts lots with the world. On the contrary, Homer (*The Iliad*, Book 15, Lines 185–92) tells us that, as a matter of fact, the gods did indeed cast lots for the world. Zeus sent his messenger Iris to order Poseidon to cease helping the Greeks during a battle in the Trojan War. Poseidon's reply was that he, Zeus and Hades, three brothers born to Kronus by Rheia, were all equal in station and they had divided the world by lot between them. Zeus drew the sky, Poseidon the sea, and Hades the underworld of the dead. Not only does God cast lots in quantum mechanics but also in molecular biology and even in arithmetic as well (Chaitin, 1987b, 1990, 1992b, 1999, 2000a and 2000b) in spite the quotation of Socrates in above.

It is important to realize that Urania, the Muse of Science, goes about her work managing the orbits of the planets and the mutations of the genome. It is up to us to find out how she does this. So if we are to understand Urania's methods we must obtain data by *measuring, counting and weighing* and introduce ideas or axioms to understand these data (Plato, *The Republic*, Book 10). But experiments are always subject error. The theory of probability enables us to find as much truth as there is in our measuremets.

What we now call the theory of probability did not become a mathematical discipline until the mid-sixteenth century with the publication of *Liber de Ludo Aleae* (The Book on Games of Chance) by Girolamo Cardano (1501–76), an Italian mathematician, physician, and gambler. Certain problems were solved by the famous mathematicians James Bernoulli (1654–1705), Daniel Bernoulli (1700–82), Pierre de Fermat (1601–65), and Blaise Pascal (1623–62). The publication of *Theorie Analytique des Probabilities* by Laplace (1796) put the subject on its first firm foundation. The hand and mind of Pierre Simon de Laplace (1749–1827) is still seen in the theory of probability.

MA1.1.2 The interpretation of probability

Before plunging into the theory, it is well to consider three important classes of interpretations of the formalism. These interpretations are not necessarily mutually exclusive and all three will be needed to address problems in this book.

(a) The classical or empirical frequency interpretation. This interpretation identifies probability directly with the empirical frequency of events. In the classical "frequentist" or statistical interpretation, probability has meaning

only in relation to an *ensemble* of trials. Suppose one has a repeatable event, A, such as obtaining a head in the tossing of a coin, drawing the ace of spades from a well-shuffled deck, finding six dots on the top of a tossed die or finding a particular value, *e.g.*, seven for the sum of the numbers showing on two tossed dice. Let a be the number of times that event A is produced in the ensemble and n be the total number of tosses or draws. Then it is asserted that as n becomes very large:

$$\frac{\lim}{n \to \infty} \frac{a}{n} = p, \tag{MA1.1}$$

where p is called the *probability* of the event. This is known as the weak law of large numbers (Feller, 1968). The *lim* is put in italics, as it is not a limit in the usual mathematical meaning of the term (Lindley, 1965). Furthermore, no law of physics prevents large deviations from Equation MA1.1. Empirical tests show merely that such deviations are so very infrequent that for practical purposes they may be neglected. Should the theory of probability stay at this intuitive interpretation many of the problems in this book could not be addressed. In particular, the several scenarios for the origin of life could not be considered, because that is not for us a repeatable event and therefore we cannot generate an ensemble of events.

The frequency definition of probability shown in the Equation MA1.1 is circular, as the idea of equally likely events is the same as equally probable events. Except in games of chance, which in their ideal form are specifically set up in this way, very few real problems have equally likely events. This difficulty of logical circularity is resolved by basing the theory of probability on a set of axioms.

Hilbert (1900, 1902) called for the development of axioms for the theory of probability in a paper that delineated his suggestions of the most important mathematical problems yet to be solved. His sixth problem suggests: "To treat in the same manner (as geometry) by means of axioms, those physical sciences, in which mathematics plays an important part; in the first rank are the Theory of Probabilities and Mechanics." This stimulated a great deal of creative activity and some of Hilbert's problems have been solved. In particular, Kolmogorov (1933) succeeded in setting up a set of axioms that solved Hilbert's sixth problem.

(b) The degree of belief or subjective interpretation. Other than in games of chance in their ideal form, most of the events with which one is confronted

and for which one must make some sort of decision in advance are not repeatable, at least they are not exactly repeatable. Predictions of the weather, the course of market prices, one's employability, and one's longevity are examples of events, the probability of which one must perforce have some knowledge in order to make the best of certain possible decisions.

A further example is the formation of probability risk assessments for large scale accidents in complicated engineering systems such as passenger and military aircraft, chemical process facilities, nuclear power plants, and their associated radioactive waste repositories. The object of probability risk assessments is to find the best administrative and engineering procedures to reduce the probability of catastrophic events below an acceptable figure. The adherents of this interpretation regard probabilities as logical weights or degrees of belief.

(c) The probable inference or inductive logic interpretation. In this interpretation we are interested in the logical support the evidence has for selected hypotheses. The hypothesis might be the guilt or innocence of a person on trial in a court of law. In such cases, it is reasonably certain that an event has occurred, that is that a crime has been committed. What must be established is the probability of the guilt or innocence of the person accused. This interpretation directs one to reject hypotheses that are believed to have a low probability in favor of ones that have an acceptably high probability.

It is important to distinguish between the mathematical theory of probability and the application to a model of the system under consideration. The model of the system is set up to reflect our best knowledge of the real world system. If the calculations prove to be incorrect, assuming that one has made no blunder in the mathematics, it is the model or its parameters that are incorrect and not the mathematical theory.

Many people, who should know better, believe that a success is "due" after a series of failures in independent trials. A baseball player who has struck out for a number of times is now believed to be "due." The manager of the team and the audience feel that he is virtually certain to get a hit. The batter can count on no help from Fortuna. Unless there has been considerable retraining, with performance demonstrated in practice, the probability for a hit in the next at-bat is not changed, *per se*, by a long sequence of strike-outs. The probability of the outcomes of experiments or trials that are independent continues to have the same value regardless of the outcomes of preceding experiments or trials.

Table MA1.1 *The probability distribution for events in the sample space of two fair dice*

Value of random variable x_i	2	3	4	5	6	7	8	9	10	11	12.
No. of elementary events of x_i	1	2	3	4	5	6	5	4	3	2	1
Probability p_i	1/36	2/36	3/36	4/36	5/36	6/36	5/36	4/36	3/36	2/36	1/36

It is worthwhile to notice at this point that the term probability as used in this book has taken on a mathematical meaning that is quantitative and not necessarily the same as that used in general conversation.

MA1.1.3 Random variables

Definition: A number x, that is defined on the elementary events A in a sample space Ω, is called a *random variable*. that is, x, the random variable, is selected by the *outcome* of an experiment. Any function, $f(x)$, is also a random variable.

Suppose, for example, we consider the roll of a die. The number on the top face and all functions of that number are random variables. The probability of each event is established in the following way. We assume each die is fair, that is, that it is an accurate unweighted cube. There are six possible outcomes, each of which can happen in only one way. Therefore we assign the probability 1/6 to the number that appears on each face. If we had prior knowledge of the behavior of the die showing that it is not an accurate cube, or has been weighted in some way, that would be reflected in the model of the system by assigning slightly different *prior* probabilities to each face.

Now consider the roll of two fair dice. The sample space is Ω^2 and there are 6^2 or 36 elementary events or outcomes. The sum of the numbers on the top face after a roll is the random variable x_i taking values from 2 to 12. Each value corresponds to an event. Any function of those numbers is also a random variable. In this case some events can happen in more than one way. For example, the sum seven can happen in six ways. We establish the probability, p_i, of each value x_i, of the random variable, by dividing the number of elementary events by which each sum can occur by the total number of elementary events in the sample space. The set of probabilities for each event in the sample space will be called the probability distribution. This is exhibited in Table MA1.1.

The probability distribution of random variables has a mean or *expectation value*. Notice that seven is both the median and the mode as the probabilities are symmetric in this case. The expectation value, $\langle x_i \rangle$, of a random variable is defined as follows:

$$\langle x_i \rangle = \Sigma_i p_i x_i. \tag{MA1.2}$$

Thus, the expectation value of any function of x_i is:

$$\langle f(x_i) \rangle = \Sigma_i p_i f(x_i). \tag{MA1.3}$$

To explain this in a simple fashion, suppose a lottery has \$1,000 to be awarded to the winner. A gambler knows, or believes, that one thousand tickets have been sold. What can he afford to pay for a ticket in order to break even in the long run? Since the probability of winning is 1/1000, he can afford to pay \$1.00.

MA2.1 Probability matrices and probability vectors

MA2.1.1 Definitions of matrices and vectors

In order to deal with coding problems and the application of the theory of probability to molecular biology we shall need the powerful concepts of *matrices* and *vectors*. As I pointed out previously (Yockey, 1974, 1992), the genetic code is represented by a matrix and the codons and amino acids by vectors. This formalism will lead us to some very important theorems that will be useful in the discussion of phylogenetic trees and evolution Michison & Durbin, 1995.

The theory of matrices and vectors comes from the properties of a set of m linear equations in n unknowns, x_j where the a_{ij} and the y_i are specified by the problem.

$$\sum_{j=1}^{n} a_{ij} x_j = y_i (i = 1, \ldots m). \tag{MA1.4}$$

Definition: A rectangular array of quantities a_{ij}, set out in m rows and n columns is called an $m \times n$ *matrix*. Matrices will be indicated in this book in **bold capital** letters. The usual convention in the literature is to let the first subscripts i refer to the rows of the matrix and the second subscripts j refer to the columns. The quantities a_{ij} are called the *elements* of the matrix.

Two matrices that have the same number of rows and columns are said to be the same *size*.

$$
\begin{bmatrix}
a_{11} & a_{12} & a_{13} & \cdots & \cdots & \cdots & \cdots & \cdots & \cdots & a_{1n} \\
a_{21} & a_{22} & \cdots & \cdots & \cdots & \cdots & \cdots & \cdots & \cdots & \cdots \\
a_{31} & \cdots & \cdots & \cdots & a_{ij} & \cdots & \cdots & \cdots & \cdots & \cdots \\
\cdots & \cdots & \cdots & \cdots & \cdots & \cdots & \cdots & \cdots & \cdots & \cdots \\
\cdots & \cdots & \cdots & \cdots & \cdots & \cdots & \cdots & \cdots & \cdots & \cdots \\
a_{m1} & \cdots & \cdots & \cdots & \cdots & \cdots & \cdots & \cdots & \cdots & a_{mn}
\end{bmatrix} = \mathbf{A}.
$$

If $m = n$ the matrix is a *square matrix* of order n. If a matrix is square, the elements where $i = j$ are called the *diagonal elements*.

Definition: Given a matrix $\mathbf{A}$ in which $m = n$, the elements of which are a_{ij} then the *determinant* of $\mathbf{A}$ written either $|\mathbf{A}|$ or $|a_{ij}|$ is, in the case where $n = 3$:

$$
\begin{vmatrix}
a_{11} & a_{12} & a_{13} \\
a_{21} & a_{22} & a_{23} \\
a_{31} & a_{32} & a_{33}
\end{vmatrix}
= \begin{aligned}
& a_{11}a_{22}a_{33} + a_{12}a_{23}a_{31} + a_{13}a_{32}a_{21} \\
& - a_{31}a_{22}a_{13} - a_{32}a_{23}a_{11} - a_{33}a_{12}a_{21}.
\end{aligned}
$$

In other cases the rule for forming these products is the same. One starts at a_{11} and forms the product of each of the n elements going diagonally down. One then begins with a_{12} multiplying together all elements, going diagonally down, together with a_{31} so that each term is the product of n elements. The last term begins with a_{1n}. Each of these terms carries a plus sign. One then begins at a_{m1} and forms the products of all elements going up along the diagonal. Then one begins again at a_{m2} multiplying together all the $n - 1$ diagonal elements with a_{11} so that each term is the product of n elements. One continues in this manner until a_{mn} is reached. Each of these terms carries a minus sign. The sum of all terms so formed is the value of the determinant $|a_{ij}|$.

Definition: Cramér's Rule

If $m = n$ and if the determinant $|a_{ij}| \neq 0$ the set of equations (MA1.4) can be solved by means of Cramér's Rule, (Gabriel Cramér, 1704–1752). This rule states that the value of x_j is given by replacing the jth column in the determinant $|a_{ij}|$ by the column y_i and dividing that determinant by $|a_{ij}|$.

If there are more than three equations this becomes tedious. However, programs are available for personal computers and even for sophisticated pocket calculators that calculate determinants and accomplish this task quite easily.

Definition: Given an m × n matrix **A**, the n × m matrix resulting from the interchange of rows and columns is called the transpose of **A**. That is, the first row of **A** becomes the first column of the transpose, the second row of **A** becomes the second column of the transpose and so forth. The usual notation for the transpose of **A** is $\mathbf{A}^T$.

Definition: A square matrix, **P**, is called a *stochastic matrix* if the elements of each of its *columns* is non-negative and their sum is equal to 1. If in addition the sum of the elements of each of its rows is equal to 1 the matrix is called *doubly stochastic* (Moran, 1986).

Some authors (Feller, 1968; Hamming, 1986), define a stochastic matrix as the transpose, $\mathbf{P}^T$, so that each of the *rows* of $\mathbf{P}^T$, is non-negative and their sum is equal to 1.

Definition: A square matrix of order n that has all elements on the diagonal equal to one and all others are equal to zero is called an identity matrix, $\mathbf{I}_n$.

Definition: An m × n matrix $\mathbf{A}(\lambda)$ whose elements are polynomials in λ is called a λ matrix. $\mathbf{A}(\lambda)$ is said to be singular or non-singular according to whether the determinant of $\mathbf{A}(\lambda)$ is zero or not.

Definition: If all the elements of a matrix are zero it is called the *null matrix* and written **O**.

Definition: Any ordered sequence of n numbers is called an ordered n-tuple. An ordered n-tuple is also called an n-dimensional vector. Each of the rows and columns of a matrix is a *vector*. In this book, vectors will be indicated by **bold lower case** letters. The numbers in the sequence are known as the *components* of the vector.

Definition: Matrices that have only one row are called *row* or $1 \times n$. A $1 \times n$ matrix is referred to as *row* vector of dimension or order n. Matrices that have only one column are called *column* or $m \times 1$ matrices. An $m \times 1$ matrix is also referred to as a *column* vector of dimension or order m. The elements are called the *components* of the vector.

A 1×1 matrix is called a scalar.

Definition: A vector that has all non-negative real number components, the sum of which is 1, is called a *stochastic vector*.

Each of the columns of stochastic matrix is a stochastic vector.

1.3.2 Algebraic properties of matrices and vectors. The mathematical power and usefulness of matrices and vectors comes largely from the fact that one may construct an algebra of these arrays that has many of the properties of the algebra of numbers. The algebra of matrices enables one to manipulate large collections of data displayed in these arrays without *ad hoc* assumptions and in a simple and convenient fashion. This avoids the usual necessity of resorting to averages. Averaging destroys information. There is a large number of powerful and useful theorems in the algebra of matrices and vectors. We shall see that many of these theorems are not intuitive and indeed some are counterintuitive. Therefore, their existence would not be suspected without the development of this algebra.

Two matrices of the same size may be added or subtracted by adding or subtracting each element with the same indices, i, j. A matrix may be multiplied by a scalar by multiplying each element by the scalar.

We are led to a natural definition for the algebraic operation of *matrix multiplication* by the following argument: Suppose we consider the set of two linear equations in three unknowns x_1, x_2, and x_3:

$$a_{11}x_1 + a_{12}x_2 + a_{13}x_3 = y_1 \tag{MA1.5}$$

$$a_{21}x_1 + a_{22}x_2 + a_{23}x_3 = y_2. \tag{MA1.6}$$

Let us make a change of variables to three equations in two unknowns y_1 and y_2:

$$b_{11}y_1 + b_{12}y_2 = z_1 \tag{MA1.7}$$

$$b_{21}y_1 + b_{22}y_2 = z_2 \tag{MA1.8}$$

$$b_{31}y_1 + b_{32}y_2 = z_3. \tag{MA1.9}$$

Upon substituting in Equations MA1.7, MA1.8, and MA1.9 for y_1 and y_2 from Equations MA1.5 and MA1.6 one has:

$$(b_{11}a_{11} + b_{12}a_{21})x_1 + (b_{11}a_{12} + b_{12}a_{22})x_2$$
$$+ (b_{11}a_{13} + b_{12}a_{23})x_3 = z_1 \tag{MA1.10}$$

$$(b_{21}a_{11} + b_{22}a_{21})x_1 + (b_{21}a_{12} + b_{22}a_{22})x_2$$
$$+ (b_{21}a_{13} + b_{22}a_{23})x_3 = z_2 \tag{MA1.11}$$

$$(b_{31}a_{11} + b_{32}a_{21})x_1 + (b_{31}a_{12} + b_{32}a_{22})x_2$$
$$+ (b_{31}a_{13} + b_{32}a_{23})x_3 = z_3. \tag{MA1.12}$$

Equations MA1.10 through MA1.12 define the algebraic operation of product of a matrix and a column vector to yield another column vector.

$$\begin{bmatrix} a_{11} & a_{12} & a_{13} \\ a_{21} & a_{22} & a_{23} \end{bmatrix} \begin{bmatrix} x_1 \\ x_2 \\ x_3 \end{bmatrix} = \begin{bmatrix} y_1 \\ y_2 \end{bmatrix} \tag{MA1.13}$$

$$\begin{bmatrix} b_{11} & b_{12} \\ b_{21} & b_{22} \\ b_{31} & b_{32} \end{bmatrix} \begin{bmatrix} y_1 \\ y_2 \end{bmatrix} = \begin{bmatrix} z_1 \\ z_2 \\ z_3 \end{bmatrix}. \tag{MA1.14}$$

In similar fashion, Equations MA1.10, MA1.11, and MA1.12 may be written in matrix notation:

$$\begin{bmatrix} b_{11}a_{11}+b_{12}a_{21} & b_{11}a_{12}+b_{12}a_{22} & b_{11}a_{13}+b_{12}a_{23} \\ b_{21}a_{11}+b_{22}a_{21} & b_{21}a_{12}+b_{21}a_{22} & b_{21}a_{13}+b_{22}a_{23} \\ b_{31}a_{11}+b_{32}a_{21} & b_{31}a_{12}+b_{22}a_{22} & b_{31}a_{13}+b_{32}a_{23} \end{bmatrix} \begin{bmatrix} x_1 \\ x_2 \\ x_3 \end{bmatrix} = \begin{bmatrix} z_1 \\ z_2 \\ z_3 \end{bmatrix}. \tag{MA1.15}$$

Equation MA1.13 may be written in the formal matrix style:

$$\mathbf{Ax} = \mathbf{y}. \tag{MA1.16}$$

The matrix **A** may thought of as *operating* on column vector **x** to change it to column vector **y**. That is, the matrix **A** can be thought of as a mapping of the points in space X into the points of the space Y. By the same token the matrix **B** may be thought of operating on the column vector **y** to change it to column vector **z** in Equation MA1.14. That is, the mapping of the points of space X on the points of space Y may be followed by another mapping of the points of space Y onto the points of space Z.

The Equation MA1.14 can be written in the compact matrix form:

$$\mathbf{By} = \mathbf{z}. \tag{MA1.17}$$

It is natural to substitute for **y** in Equation MA1.16 from Equation MA1.17:

$$\mathbf{BAx} = \mathbf{Cx} = \mathbf{z}, \tag{MA1.18}$$

where **B A** = **C** and the elements of **C** are given in Equation MA1.15.

One notices in Equation MA1.15 that the elements of **C** are the sums of the arithmetical products of the elements in the rows of **B** and the elements in the columns of **A**. This may be shown to be true in general and forms the basis of a definition of matrix multiplication. Write Equation MA1.17 as a set of m linear equations in p unknowns:

$$\sum_{k=1}^{p} b_{ik} y_k = z_i \quad (i = 1, \ldots m). \tag{MA1.19}$$

Equation MA1.4 may be rewritten using k rather than i as a suffix in order not to confuse the suffix in Equation MA1.18. We may substitute for the y_k in Equation MA1.18 from Equation MA1.12:

$$\sum_{k=1}^{p} b_{ik} \left(\sum_{j=1}^{n} a_{kj} x_j \right) = z_i \quad (i = 1, \ldots \ldots m). \tag{MA1.20}$$

Equation MA1.20 may be rewritten:

$$\sum_{j=1}^{n} \sum_{k=1}^{p} b_{ik} a_{kj} x_j = z_j \quad (i = 1, \ldots \ldots m). \tag{MA1.21}$$

Therefore, successive mappings $\mathbf{A}x = y$ and $\mathbf{B}y = z$ yield a mapping $\mathbf{C}x = z$ where $\mathbf{C}$ is an $m \times n$ matrix with the elements c_{ij}:

$$c_{ij} = \sum_{k=1}^{p} b_{ik} a_{kj} \quad (i = 1, \ldots \ldots m) \text{ and } (j = 1, \ldots \ldots n). \tag{MA1.22}$$

Definition: The product matrix $\mathbf{C}$, is called the *Cayley product* (Arthur Cayley, 1821–95). Two matrices $\mathbf{B}$ and $\mathbf{A}$ may be multiplied if $\mathbf{B}$ is $m \times p$ and $\mathbf{A}$ is $p \times n$ where m, p and n are positive real numbers. It is easy to see, in the simple case of Equation MA1.15 and more generally in Equation MA1.22, that unless the number of rows in $\mathbf{B}$ is the same as the number of columns in $\mathbf{A}$ the product $\mathbf{BA}$ is not defined.

Definition: Two matrices that may be multiplied are said to be *conformable*. When two matrices do not meet these conditions the product is not defined, and they are said to be *non-conformable*.

The multiplication of matrices is not commutative in general. Even if $\mathbf{AB}$ are conformable, $\mathbf{BA}$ may not be. For example if $\mathbf{A}$ is 2×3 and $\mathbf{B}$ is 5×2 then $\mathbf{BA}$ is conformable but $\mathbf{AB}$ is not. If $\mathbf{A}\,\mathbf{B} = \mathbf{O}$ it is not possible to assume in general that either $\mathbf{A} = \mathbf{O}$ or that $\mathbf{B} = \mathbf{O}$. Multiplication is, however, associative if the matrices are conformable in sequence.

$$(\mathbf{A}\,\mathbf{B})\,\mathbf{C} = \mathbf{A}\,(\mathbf{B}\mathbf{C}). \tag{MA1.23}$$

Column vectors are $m \times 1$ matrices. Row vectors are $1 \times n$ matrices. They have two kinds of products. The first is called the *inner, scalar* or *dot product*. It is the product of a $1 \times n$ matrix by an $n \times 1$ by a matrix that is a 1×1 matrix and therefore a scalar. It is defined as follows:

$$\mathbf{x} \bullet \mathbf{y} = \sum_{i=1}^{n} x_i y_i. \tag{MA1.24}$$

If the inner product of two vectors vanishes they are said to be orthogonal. From Equation MA1.22 one may see that the elements c_{ij} of $\mathbf{C}$ in Equation MA1.22 are the inner products of the ith row vector of $\mathbf{B}$ and the jth column vector of $\mathbf{A}$.

The second vector product is the matrix product of an $n \times 1$ matrix by a $1 \times n$ matrix and is therefore an $n \times n$ matrix. It is called the *vector product*.

Two vectors may be added or subtracted by adding or subtracting the respective components. A vector may be multiplied by a scalar by multiplying each component by the scalar.

Theorem 1.4 The product of two stochastic matrices is also a stochastic matrix.

The proof on this theorem can be done in a few lines (Hamming, 1986).

Square matrices may be raised to positive integer powers. A stochastic matrix of any power is also a stochastic matrix. The usual laws of exponents hold.

$$\mathbf{A}^{\alpha}\mathbf{A}^{\beta} = \mathbf{A}^{\alpha+\beta}; \quad (\mathbf{A}^{\alpha})^{\beta} = \mathbf{A}^{\alpha\beta}. \tag{MA1.25}$$

Some matrices have a square root. If $\mathbf{A}^2 = \mathbf{B}$ then $\mathbf{B}^{1/2} = \mathbf{A}$. In some cases other fractional roots and exponents exist. Some matrices have no square root, others have an infinite number (Eves, 1966). $\mathbf{A}^0$ is taken to be the identity matrix, $\mathbf{I}_n$.

Definition: Given a sample space Ω with events A_i where A_i occurs with probability $p(A_i)$, then the ordered sequence $\{p(A_1), p(A_2), \ldots .p(A_n)\}$ is called a *probability vector* $\mathbf{p}$. The probability space may be described by $(\Omega, A, \mathbf{p})$.

For an evolving system, such as one undergoing mutations, the different possible states of the system may be described as different events, and at a given time the probability vector for the various states or events maybe specified. As the system evolves further, the probability vector will change, in general.

Definition: A matrix **A**, which changes one vector to another by multiplication, is called a *transition matrix* by mathematicians and theoretical physicists.

In this book, we are concerned only with *transition probability matrices* that change one probability vector to another by multiplication. The word *transition* has assumed a special meaning in molecular biology. It means a nucleotide change that does not change the purine-pyrimidine orientation. A mutation that changes purine-pyrimidine to pyrimidine-purine or pyrimidine-purine to purine-pyrimidine is called a *transversion*. In this book, to avoid the awkward terminology *transition/transversion*, a matrix that changes one probability vector to another will be called a *probability transition matrix*. The meaning will be clear depending on whether the context is mutations or matrices.

Definition: A set of events is called a *cylinder set* or, for short, a *cylinder* (Feller, 1968; Khinchin, 1957) if pairs of events each satisfy restrictive conditions. For example, consider the set of points in a Euclidean space of three dimensions, which lie within a square satisfying the conditions $0 < x < 1$ and $0 < y < 1$. If the values of z are unrestricted the points enclosed in the (x, y, z) space, with a square cross section, define a *cylinder* in that space.

The third position in eight of the familiar triplet codons in the set space G^3 may vary indefinitely among the four nucleotides in either the DNA or RNA alphabets without change of the read-off amino acid. The specificity of these codons is defined by the first two nucleotides. Such sequences meet the definition of a *cylinder set*. I propose to call these codons *cylinder codons*.

Definition: Cramér's Rule

If $m = n$ and if the determinant $|a_{ij}| \neq 0$, the set of equations can be solved by means of Cramér's Rule, (Gabriel Cramér, 1704–52). This rule states that the value of x_j is given by replacing the jth column in the determinant $|a_{ij}|$ by the column y_i and dividing that determinant by $|a_{ij}|$.

If there are more than three equations this becomes tedious. However, programs are available for personal computers and even for sophisticated pocket calculators that calculate determinants and accomplish this task quite easily.

Markov process and the random walk. Phylogenetic chains that relate one protein to another by means of mutational steps are examples of the general mathematical theory of the transition between successive states of a system. The sequence of states (or events) relating the transition between the initial and final states of a system is known as a Markov process or a Markov chain after the mathematician Andrei Andreevich Markov (1856–1922). In general, the probability of the final state is governed by the probabilities of the finite sequence of states that precede the final state. In molecular biology, we are concerned only with discrete and finite Markov chains (Kemeny, Snell, and Knapp, 1976). If there are k such states the Markov process is known as a kth-order Markov process. In information theory, where one is interested in the generation of a sequence of symbols, known as an alphabet, that generate a genetic message, a kth-order Markov process is known as a *k-memory* source.

Markov processes are an example of the fact that the probability of an event is often strongly influenced by events occurring previously. In general, several previous events may affect the conditional transition probability. Two states are said to *communicate* if one can be reached from the other. If a state is such that once entered it cannot be exited, it is called an *absorbing state*. A state through which a system may move is called a *transient state*. A *reflecting* state is one that sends the system into another state without being occupied. In molecular biology, the nucleotides in DNA or mRNA sequences and, in the genetic code, codons may be regarded as Markov states. All codons communicate, even the stop codons, by means of a sequence of single base interchanges, that is, by a Markov chain. The stop codons are usually absorbing states, but they may be transient states.

The elements of the transition probability matrix of Markov states may change as the process proceeds from one step to the next. If the elements of the transition probability matrix do not change, the Markov process is called homogeneous or *stationary*. Bernoulli processes or Bernoulli trials in which succeeding events are independent are simple examples of stationary Markov processes. The sequences so produced are called Bernoulli sequences (Solomonoff, 1964). The sequence of outcomes of the toss of a coin is a Bernoulli sequence because there are only two outcomes, the trials are independent and the probability distribution is stationary. The sequence of the numbers generated by a sequence of throws of a die is not a Bernoulli sequence since there are six outcomes.

Among the examples of the Markov state process, for which we shall have use, is one known as a *random walk*. Suppose a man is standing on the mid-field line of a football field. He flips a coin to determine which way to go, heads, one way, tails the other. The total distance moved in one direction or the other is a random variable. A number of questions can be asked. For example, what is the probability of crossing a goal line? This an example of a one-dimensional random walk. The scenario may be elaborated to more than one dimension. The man might be on the corner of two city streets. He has forgotten where his hotel is. It is very late at night and there is no one to ask. Here the Markov state system is the pair of coordinates giving the man's position on the city map. He goes one block at a time and at each corner he flips a coin twice, to determine which of the four ways to go. He starts out with the following code: HT = North, HH = South, TH = West, TT = East. All streets going in the North direction end at a river, which is an absorbing state. If he goes that far, he will fall in. There is a wall on the South end of the North–South streets, which is a reflecting state. Of course, if he does find his hotel he goes in and so the hotel is also an absorbing state. All other states are transient states. Suppose his wife is at the hotel and she goes out to look for him. She would need the probability distribution so she could go to the more probable street intersections first. He might forget his code between corners. This change in code is a mutation so that, perhaps, HH = North and HT = South. There are many other questions that could be asked and such problems may get quite complicated.

There are many applications of the random walk in science. Perhaps the most famous is the solution to the problem of Brownian motion. This was solved by Einstein (1905, 1906) and proved to be the final nail in the coffin of the continuous matter theory. After millennia of philosophical and scientific arguments, atoms indeed did exist!

Proofs of theorems in the text

Proof of the Shannon–McMillan–Breiman Theorem

(a) See Section 4.1. The Markov process is ergodic, therefore the statistical properties of the sequence are stationary. The probability of any sequence does not depend on the starting point. Suppose we start at E_{k_1}, which has probability P_{k_1}. Let i and l be two arbitrary numbers each ranging from 1 to n and let m_{il} be the numbers of pairs of the form $k_r \, k_{r+1}$ where $1 \leq r \leq N - 1$ in which $k_r = i$ and $k_{r+1} = l$. Then $p(l \mid i)$ is the transition probability from state i to state l. The values of the k's run from one to n at each step. The

subscripts indicate the position in the sequence and run from 1 to N. Then we have for m_{il} pairs:

$$p(C) = P_{k_1} \prod_{i=1}^{n} \prod_{l=1}^{n} p(l \mid i)^{m_{il}}. \qquad \text{(MA1.26)}$$

We now assign the sequence to the first group if it has the following two properties:

1. It is a possible outcome, that is, all $p(l \mid i) > 0$. For a DNA sequence that means that there are no terminator codons.
2. For any i, l the following inequality holds:

$$|m_{il} - NP_i p(l \mid i)| < N\delta. \qquad \text{(MA1.27)}$$

The second term in the absolute value bars is the estimate of m_{il} from the law of large numbers. In order to make Relation MA1.27 an equality let us find a number h_{il} the absolute value of which is less than one for any pair m_{il} and write as follows:

$$m_{il} = NP_i + N\delta h_{il}. \qquad \text{(MA1.28)}$$

Substitute the expression for m_{il} in Equation MA1.26 then take the logarithm to base 2.

$$-\log_2 P(C) = -\log_2 P_{k_1} - N\Sigma_i \Sigma_l P_i p(l \mid i) \log_2 p(l \mid i)$$
$$- N\delta\Sigma_i \Sigma_\lambda hil \, p(l/i). \qquad \text{(MA1.29)}$$

The second term on the right is simply N times the one step entropy in the chain, as we have seen above. Equation MA1.29 may be rewritten in the following form, returning to the inequality by replacing all h_{il} by 1:

$$|(-\log_2 p(C)/N) - H| < -(1/N)\log_2 P_{k_1} - \delta\Sigma_i \Sigma_l \log_2 p(l \mid i). \qquad \text{(MA1.30)}$$

The left-hand side of the inequality is less than $\eta > 0$, which is as small as we please by choosing N sufficiently large. By condition (2), we have chosen δ sufficiently small. This proves the first part of the theorem and we see that the procedure leading to Equations 4.4 and 4.5 of Chapter 4 is justified.

To prove the second part of the theorem we must find the sum of the probabilities of those sequences that do not satisfy the inequality of MA1.27.

That is:

$$\sum_{i=1}^{n}\sum_{l=1}^{n} P\{\,|\,m_{il} - NP_i\,p(l\,|\,i)| \geq N\delta\}. \qquad (MA1.31)$$

Let us start by selecting any pair i, l. By the law of large numbers we can say that the probability that $|\,m_i - NP_i\,|$ is less than $N\delta$ is less than $1 - \varepsilon$. In mathematical symbolism this is written as follows:

$$P\{\,|\,m_i - NP_i\,| < N\delta\} > 1 - \varepsilon, \qquad (MA1.32)$$

where m_i is the frequency of events in state i.

By the same token if N is sufficiently large, we can write:

$$P\{\,|\,(m_{il}/m_{i)} - p(l\,|\,i)| < \delta\} > 1 - \varepsilon. \qquad (MA1.33)$$

Therefore, the probability of satisfying both of the inequalities in the brackets is the product of the separate probabilities and is greater than $(1 - \varepsilon)^2 > 1 - 2\varepsilon$. If we multiply the inequality of MA1.32 through by $p(l\,|\,i)$ it follows that, since $1 > p(l\,|\,i) > 0$:

$$|\,p(l\,|\,i)\,m_i - NP_i\,p(l\,|\,i)\,| < p(l\,|\,i)\,N\delta < N\delta. \qquad (MA1.34)$$

If we add this inequality to the inequality that comes from Expression MA1.32, thus:

$$|\,m_{il} - m_i\,p(l\,|\,i)| < \delta m_i \leq N\delta, \qquad (MA1.35)$$

we have:

$$|\,m_{il} - NP_i\,p(l\,|\,i)\,| < 2N\delta. \qquad (MA1.36)$$

Thus, for any i and l and for N sufficiently large we have the probability that:

$$P\{\,|\,m_{il} - NP_i\,p(l\,|\,i)\,| < 2N\delta\} > 1 - 2\varepsilon, \qquad (MA1.37)$$

which is the same statement that:

$$P\{\,|\,m_{il} - NPip(l\,|\,i)| > 2N\delta\} < 2\varepsilon. \qquad (MA1.38)$$

We can now carry out the summation in Expression MA1.38 and find:

$$\sum_{i=1}^{n}\sum_{l=1}^{n} P\{\,|\,m_{il} - NPp(l\,|\,i)| > 2N\delta\} < 2n^2\varepsilon. \qquad (MA1.39)$$

Because the right side of this inequality can be made as small as we please by choosing ε sufficiently small, the sum of the probabilities of all the sequences in the second group can be made as small as we please by choosing N sufficiently large. This meets the requirement of the second part of the theorem and completes the proof.

Theorem 5.1 The value of the mutual entropy is symmetric between the source and the receiver $I(A; B) = I(B; A)$.

Proof: First express the two entropies in terms of the probabilities where $p(i, j)$ is the probability of the pair (i, j).

$$I(A; B) = -\Sigma_i p_i \log_2 p_i + \Sigma_{i,j} p(i, j) \log_2 p(i \mid j) \qquad \text{(MA1.40)}$$

$$p_i = \Sigma_j p(i \mid j). \qquad \text{(MA1.41)}$$

Substitute p_i in the first term of equation (but not in the logarithm).

$$I(A; B) = -\Sigma_{ij} p(i, j) \log_2 p_i + \Sigma_{i,j} p(i, j) \log_2 p(i \mid j)$$
$$\qquad \text{(MA1.42)}$$

$$I(A; B) = \Sigma_{i,j} p(i, j) \log_2 [p(i \mid j)/p_i]. \qquad \text{(MA1.43)}$$

The probability of the pair (i, j) is:

$$p(i, j) = p_i p(j \mid i) = p_j p(i \mid j). \qquad \text{(MA1.44)}$$

Substitute from Equation 5.8 to the argument of the logarithm in Equation 5.7:

$$I(A; B) = \Sigma_{i,j} p(i, j) \log_2 [p(j \mid i)/p_j] = I(B; A), \qquad \text{(MA1.45)}$$

which proves the theorem.

Theorem 5.2 The mutual information $I(A; B)$ is zero, if and only if, the sequences in alphabet A and those in alphabet B are independent.

Proof: From Equation MA1.44 we have, assuming that $p_j \neq 0$.

$$p(i, j)/p_j = p(i \mid j). \qquad \text{(MA1.46)}$$

Substitute $p(i \mid j)$ in the argument of the logarithm in Equation MA1.43.

$$I(A; B) = -\Sigma_{i,j} p(i, j) \log_2 \left[\frac{p(i \mid j)}{p_i p_j} \right]. \qquad \text{(MA1.47)}$$

The sequences in alphabet A and and those in alphabet B are independent *if and only if* $p(i \mid j) = p_i p_j$. *In that case the argument of the logarithm in equation MA1.43 is unity and the logarithm vanishes for all pairs* (i, j).

(We recall that a regular matrix is one in which, at some power, all matrix elements are > 0.)

Let us remind ourselves that $\mathbf{P}^t$ is a λ-matrix where the elements are polynomials of degree t. We can, if we wish, stop at any step t to calculate the matrix elements and vector components for substitution in Equations 5.13 and 5.14. We can, as a matter of fact, follow the progress of the mutual entropy, step by step, to its equilibrium value.

The Perron–Frobenius Theorem 1.5 (Bellman, 1997; Berman and Plemmons, 1994; Frobenius, 1908, 1912; Lancaster and Tismenetsky, 1985; Marcus and Minc, 1984; Perron, 1907; Petersen, 1983; Seneta, 1973):

The Perron–Frobenius Theorem is not intuitive, so that, if we attempt to find an equilibrium fixed probability vector by raising $\mathbf{P}$ to higher powers, we are in for a formidable amount of computation. However, the mathematics leads us to the correct result with unseemly ease and, in particular in the last step, reminds us that α must not be equal to zero.

Let $\mathbf{A}$ be a regular stochastic matrix. Then:

    a. A fixed probability vector $\mathbf{t}$, none of whose components is zero, is associated with $\mathbf{A}$.

    b. The sequences of powers $\mathbf{A}, \mathbf{A}^2, \mathbf{A}^3, \ldots$ approaches the matrix $\mathbf{T}$ whose rows are each the vector $\mathbf{t}$.

    c. If $\mathbf{p}$ is any probability vector, then the sequence of vectors, $\mathbf{Ap}, \mathbf{A}^2\mathbf{p}, \mathbf{A}^3\mathbf{p} \ldots$ approaches the fixed probability vector $\mathbf{t}$.

Proof: Because of the normalization condition $\Sigma_i p_{ij} = 1$, we have:

$$\max_{ij} p_{ij} \le 1 - (n - 1)\delta, \tag{MA1.48}$$

where

$$\delta = \min_{ij} p_{ij}.$$

Consider the equation:

$$p_{ij}^{(t+1)} = \Sigma_k p_{ik}^{(t)} p_{kj}. \tag{MA1.49}$$

Let

$$m_i^{(t)} = \min_j \, p_{ij}^{(t)}$$
$$M_i^{(t)} = \max_j \, p_{ij}^{(t)}$$

$$m_i^{(t+1)} = \min_j \Sigma_k p_{ik}^{(t)}\, p_{kj}$$
$$m_i^{(t=1)} \geq mi\,(t)\,\min_j \Sigma_k p_{kj}$$
$$m_i^{(t=1)} \geq m_i^{(t)}.$$

Thus the sequence $m_i^{(1)}, m_i^{(2)}, \ldots$ is non-decreasing, and by the same token, $M_i^{(1)}, M_i^{(2)} \ldots$ is non-increasing. Therefore, both sequences tend to a limit. We now prove these limits to be equal. Some terms in the equation: $\Sigma_i(p_{ij} - p_{ik}) = 0$ for a given j and k, will be such that $p_{ij} \geq p_{ik}$ and some may be such that $p_{ij} < p_{ik}$. Let Σ^+ indicate summation of the first set and Σ^- summation over the second set. Then the two sets may be arranged in this manner:

$$\Sigma_i^+(p_{ij} - p_{ik}) = \Sigma_i^+ p_{ij} - \Sigma_i^+ p_{ik} \tag{MA1.50}$$
$$\Sigma_i^+(p_{ij} - p_{ik}) = 1 - \Sigma_i^- p_{ij} - \Sigma_i^+ p_{ik}. \tag{MA1.51}$$

Let s be the number of values of i in the first set,

$$\Sigma_i^+(p_{ij} - p_{ik}) \leq 1 - (n-s)\delta - s\delta \tag{MA1.52}$$
$$\Sigma_i^+(p_{ij} - p_{ik}) \leq 1 - n\delta.$$

This leads to:

$$M^{(t+1)} - m_l^{(t+1)} = \max_{jk}\Sigma_i\, p_{li}^{(t)}(p_{ij} - p_{ik}) \tag{MA1.53}$$

$$M^{(t+1)} - m_l^{(t+1)} \leq \max_{jk}\{\Sigma_i^+ p_{li}^{(t)}(p_{ij} - p_{ik})$$
$$+\Sigma_i^- p_{li}^{(t)}(p_{ij} - p_{ik})\} \tag{MA1.54}$$

$$M^{(t+1)} - m_l^{(t+1)} \leq \max_{jk}\{\Sigma_i^+ M_l^{(t)}(P_{ij} - p_{ik})$$
$$+\Sigma_i^- m_l^{(t)}(p_{ij} - p_{ik})\} \tag{MA1.55}$$

$$M^{(t+1)} - m_l^{(t+1)} \leq \max_{jk}\Sigma_i^+\left(M_l^{(t)} - m_l^{(t)}\right)(p_{ij} - p_{ik}) \tag{MA1.56}$$

$$M^{(t+1)} - m_l^{(t+1)} \leq (1 - n\delta)\left(M_l^{(t)} - m_l^{(t)}\right). \tag{MA1.57}$$

We can now conclude that:

$$M_l^{(t)} - m_l^{(t)} \leq (1 - n\delta)^t. \tag{MA1.58}$$

Therefore, $M_l^{(t)}$ and $m_l^{(t)}$ approach the same non-zero constant as t increases. If the probability matrix is doubly stochastic and the vector $\mathbf{p}$ satisfies the equation $\mathbf{pP} = \mathbf{p}$, then each component of $\mathbf{p}$ is $1/n$.

Definition: An $m \times n$ matrix $\mathbf{A}$ (λ) whose elements are polynomials in λ is called a λ matrix. $\mathbf{A}$ (λ) is said to be *singular* or *non-singular* according to whether the determinant $|\mathbf{A}(\lambda)|$ is zero or not.

MA2.1 The role of axioms in mathematics

Since before the time of Euclid of Alexandria (325 B.C.–265 B.C.), mathematicians have preferred to treat their subject in terms of *axioms*. This avoids an endless sequence of contrived and *ad hoc* assumptions being made in pursuit of the desired result. It prevents us from arguing a case, as it were, to achieve a foreordained result. The axiomatic treatment often reveals unsuspected theorems that are not intuitive. It discloses relations between problems that would otherwise not be realized. In particular, the axiomatic treatment of the theory of probability avoids logical circularities and allows the application of the theory to a broad range of problems.

Axioms are elementary facts that cannot be explained by reducing them to simpler ones, rather, they must be taken as a starting point. Axioms are not necessarily self-evident or representative of the real world. It is only in the application to real-world problems that they are useful or useless, appropriate or inappropriate, interesting or dull, to the degree of the exactness that the set of axioms is a mathematical representation of the real world.

A set of axioms, or postulates if one prefers, must be *consistent*; that is, it must not be possible to prove a given statement or theorem both true and false. Axioms must be *independent*; that is, any axiom that can be proved from the others is a theorem and must be crossed off the list. The set of axioms must be *unique* and *finite* in number. The axioms must be *complete*; that is, one must not need to introduce *ad hoc* statements as one goes about proving theorems. Gödel (1931) that proved that for any set of axioms there are true statements that cannot be proved from the axioms. Furthermore, there are questions that are undecidable from the set of axioms. Reasoning from axioms is the highest form of human thought. Nevertheless, there are questions that are beyond human reasoning (Chaitin, 1987a, Ch. 11).

Glossary

algorithm   A step-by-step problem-solving procedure, especially an established, recursive computational procedure for solving a problem in a finite number of steps.

amino acid   Any of various organic compounds charactrized by the presence of the amino group ($-NH_2$) and a carboxylic group ($-COOH$).

axiom   Axioms are elementary facts that cannot be explained by reducing them to simpler ones; rather, they must be taken as a starting point.

bases   Any of the combinations of three nucleotides in DNA and RNA.

biochemistry   The study of the chemistry of living organisms. See comparison with organic chemistry.

bit   The information in the binary source alphabet is called a *bit*.

block code   A code in which all the code letters have the same number of elements.

byte   The number of bits in a computer code or the genetic code extended from a binary source alphabet is called *byte*. A byte is not always eight bits – the number of bits in the byte depends on the code.

catalysis   Modification or an increase in rate of chemical reactions induced by a chemical that is unchanged at the end of the reaction.

chiral   Objects or molecules that are formed in both right-handed and left-handed condition. They are mirror images of each other.

chromosome   A thread-like body in the nucleus of an organism that contains the genes that are composed of DNA and protein.

code Given a source with probability space $[\Omega, A, \mathbf{p_A}]$ and a receiver with probability space $[\Omega, B, \mathbf{p_B}]$, then a unique mapping of the letters of alphabet A on to the letters of alphabet B is called a *code* (Perlwitz, Burks, and Waterman, 1988).

codon A sequence of three nucleotides in DNA or mRNA that specifies a particular amino acid during protein synthesis. Of the sixty-four possible codons three are stop codons that do not usually specify an amino acid.

cytochrome c A small globular heme, containing an iron ion, formed early in the evolution of life

cytosine One of the five nitrogen containing bases that compose nucleic acids.

dialectical materialism The philosopohical belief that the appearance of life is achieved, not through the laws of physics and chemistry, but through the *Law of the Transformation of Quantity into Quality*.

digital Elements that form a signal, message, or sequence.

DNA (deoxribose nucleic acid) A long linear sequence of four kinds of deoxyribose nucleotides that carry the genetic information. In its native state, DNA is a double helix of two antiparallel strands held together by hydrogen bonds between complementary purine and pyrimidine bases.

enzyme Any of numerous complex proteins that catalyze specific bio-chemical reactions.

evolution The theory that groups of organisms change with passage of time, mainly as a result of natural selection, so that descendants differ morphologically and physiologically from their ancestors.

gene That segment of the genome that contains the genetic message for a specific protein.

genetic code The table of correspondence between the codons in DNA and amino aicds.

genetic message The sequence of codons that make up DNA and contain the information that controls the formation of protein.

genome The total genetic encyclopedia of genetic messages in an organism.

Hamming distance The number of positions in which synonymous code words differ is called the Hamming distance (Hamming, 1950).

information content The number of bits or bytes in a message or sequence of letters selected from an alphabet.

isotope   Any of two or more chemical element having the same electrical charge but a different atomic mass.

majority logic redundancy   Sending the message several times to overcome errors.

materialist–reductionist   One who believes that life processes or mental acts can be completely explained by chemical and physical laws.

matrix   A rectangular array of quantities a_{ij}, set out in m rows and n columns is called an $m \times n$ *matrix*. See the Mathematical Appendix.

mechanism–reductionism   The doctrine that all natural phenomena are explicable by material causes and mechanical principles.

nitrogen fixation   The assimilation of atmospheric nitrogen by living organism.

nucleic acid   Any of several organic acids formed of a sugar (deoxyribose or ribose) with attached purine (adenine and/or guanine) and pyrimidine, nitrogen containing bases.

nucleoside   A small molecule composed of a purine or a pyrimidine base linked to a pentose (either ribose or deoxyribose).

nucleotide   A nucleoside with one or more phosphate groups liked via an ester bond to the sugar moiety. DNA and RNA are polymers of nucleotides.

ontogeny   The growth and development of an organism from fertilization to sexual maturity and senescence.

optical isomers   Molecules that are mirror images of each other.

organic chemistry   The chemistry of carbon compounds.

organic compound   A chemical compound that may be composed of carbon, hydrogen, nitrogen, and oxygen.

phylogeny   The evolutionary history of an organism.

polypeptide   A molecule composed of three or more amino acids joined in a chain.

postulate   A synonym for axiom.

probability sample space   See the Mathematical Appendix.

probability theory   See the Mathematical Appendix.

protein   One of a large group of biomolecules that are composed of chains of amino acids. Some contain metals such as iron, zinc, copper, and manganese. Thyroxine contains iodine.

proteome   The collection of proteins in molecular biology.

racemic Substances composed of equal amounts of molecules of a right- and left-handed form.

reductionist One who believes that life processes or mental acts are instances of chemical and physical laws.

ribsome The RNA-rich cytoplasmic granules that are the sites of protein synthesis.

sense code letters Code letters that have been assigned a meaning or significance.

sense versus non-sense Meaning contrasted with unassigned significance.

sequence hypothesis The linear and digital sequence of nucleotides in the genome, composed of DNA, that contains the genetic information that controls the formation of proteins.

specificity Having a definite meaning or significance.

speculation Reasoning based on inconclusive evidence.

spontaneous generation The presumed origin of life from non-living organic compounds.

stochastic A chance process, that is, probability. See the Mathematical Appendix.

sugar Any of the oligosaccharides, such as sucrose, fructose, having a generalized chemical formula CH_2O.

theory A general principle or sets of principles that describe process or conditions in nature that is supported by a large body of evidence and has been repeatably tested by experiment and has been found to be applicable to a large variety of circumstances. See also **speculation**.

transcription The decoding of the genetic message from the DNA alphabet to the mRNA alphabet is called *transcription*.

translation The genetic message is decoded by the ribosomes from the sixty-four-letter mRNA alphabet to the twenty-letter alphabet of the proteome. This decoding process is called *translation* in molecular biology.

tRNA (transfer RNA) A group of small RNA molecules that function as amino acid donors during protein synthesis.

uracil One of the five nitrogen-containing bases present in biological nucleic acids.

Urschleim (primeval slime) The colloids or coacervates generated from organic substances in the early ocean from which Ernst H.P.A. Haeckel (1834–1919) claimed life originated by self-organizing biochemical cycles.

virus A noncellular infectious particle composed of nucleic acid and surrounded by a protein coat.

vitalism The belief that all matter possesses a *World Spirit* and organized bodies, especially living organisms have it to an intense degree.

References

Abelson, P. H. (1966). Chemical events on the primitive Earth. *Proceedings of the National Academy of Sciences*, **55**, 1365–72.

Adami, C., & Cerf, N. J. (2000). Physical complexity of symbolic sequences. *Physica D*, **137**, 62–9.

Adami, Christoph, Ofria, Charles, & Collier, Travis. (2000). Evolution of biological complexity. *Proceedings of the National Academy of Sciences*, **97**, 4463–668.

Adams, Keith L., Cronn, Percifield, Ryan, & Wendel, Jonathan F. (2003). Genes duplicated by polyploidy show unequal contributions to the transcriptome and organ-specific reciprocal silencing. *Proceedings of the National Academy of Sciences*, **100**, 4649–54.

Adelman, J. P., Bond, C. T., Douglas, J., & Herbert, E. (1987). Two mammalian genes transcribe from opposite strand of the same DNA. *Science*, **235**, 1514–17.

Aerssens, Konings Frank, Luyten Walter, Macciardi Fabio, Sham Pak C., Straub Richard E., Weinberger Daniel R., Cohen Nadine, & Cohen Daniel. (2002). Genetic and physiological data implicating the new human gene G72 and the gene for D-amino acid oxidase in schizophrenia. *Proceedings of the National Academy of Sciences*, **99**, 13675–80.

Aklerlof, G. C., & Wills, E. (1951). Bibliography of Chemical Reactions in Electrical Discharges. Project NR223-064. Office of Technical Services, Department of Commerce, Washington DC.

Alber, T., Bell, J. A., Dao-pin, S., Nicholson, H., Wozniak, J. A., Cook, S., & Matthews, B. W. (1987). Contributions of hydrogen bonds of Thr[157] to thermodynamic stability of phage T4 lysozyme. *Nature*, **330**, 41–6.

Alber, T., Bell, J. A., Dao-pin, S., Nicholson, H., Wozniak, J. A., Cook, S., & Matthews, B. W. (1988). Replacements of Pro[86] in T4 lysozyme extend an α-helix but do not alter protein stability. *Science*, **239**, 631–5.

Almasy, R. J., & Dickerson, R. E. (1978). *Pseudomonas* cytochrome c_{551} at 2 Å resolution: Enlargement of the cytochrome c family: *Proceedings of the National Academy of Sciences*, **75**, 2674–8.

Altschul, S. F., et al. (1997). Gapped BLAST and PSI-BLAST: A new generation of protein database search programs. *Nucleic Acid Research*, **25**, 3389–402.

219

Amelin, Yuri, Krot, Alexander N., Hutcheon, Ian D., & Ulyanov, Alexander A. (2002). Lead isotopic ages of chondrules and calcium-aluminum inclusions. *Science*, **297**, 1678–83.

Amis, Martin. (2002). *Koba the Dread and the Twenty Million*, Vintage Books 6613, E. Mill Plain Blvd. Vancouver, WA 98661, USA.

Anderson, I. G., de Bruijn, M.H.L., Coulson, A. R., Eperon, O. C., Sanger, F., & Young, I. G. (1982). Complete sequence of bovine mitochondrial DNA. *Journal of Molecular Biology*, **156**, 683–717.

Anderson S., Bankier, A. T., Barrell, B. G., de Bruijn, M. H. L., Coulson, A. R., Drouin, J., Eperon, I. C., Nierlich, D. P., Roe, B. A., Sanger, F., Schreier, P. H., & Young, I. G. (1981). Sequence and organization of the human mitochondrial genome. *Nature*, **290**, 457–65.

Anhäuser, Marcus. (2003). Baumeister des Lebens, *Süddeutsche Zeitung*, May 13, 2003.

Applebaum, Anne. (2003). *GULAG. A History*. New York: Doubleday.

Archibald, John M., Rogers, Matthew B., Toop, Michael, Ishida, Ken-ichiro & Keeling, Patrick J. (2003). Lateral gene transfer and the evolution of plastid-targeted proteins in the secondary plastid-containing alga *Bigelowiella natans*. *Proceedings of the National Academy of Sciences*, **100**, 7678–83.

Arrhenius, S. (1908). *Worlds in the Making*. London: Harper.

Ash, Robert B. (1965). *Information Theory*. New York: Dover Publications.

Astrobiology.com Press Release Thursday, August 21, 2003, New findings could dash hopes for past oceans on Mars. www.astrobiology.are.nasa.gov/

Avery, Oswald T., MaCleod, Colin M., & McCarty, Maclyn. (1953). Studies on the chemical nature of the substance inducing transformation of pneumococcal types. *Journal of Experimental Medicine*, **79**, 137–59.

Baba, M. L., Darga, L. L., Goodman, M., & Czelusniak, J. (1981). Evolution of cytochrome c investigated by maximum parsimony method. *Journal of Molecular Evolution*, **17**, 476–7.

Bada, Jeffrey, L. (1997). Enhanced: Extraterrestrial Handedness? *Science*, **275**, 942–3.

Bada, Jeffrey, & Lazcano, Antonio. (2002a). Some like it hot, but not the first biomolecules. *Science*, **296**, 1982–3.

Bada, Jeffrey, & Lazcano, Antonio. (2002b). Miller revealed new ways to study the origins of life. *Nature*, **416**, 475.

Bada, Jeffrey L., & Lazcano, Antonio. (2003). Prebiotic Soup–Revisiting the Miller Experiment, *Science*, **300**, 745–6.

Bada, Jeffrey, L., Glavin, Daniel, P., McDonald, Gene D., & Becker, Luann. (1998). A search for endogenous amino acids in Martian meteorite ALH84001. *Science*, **279**, 362–5.

Balasubramaian, Suganthi, Schneider, Tamara, Gerstein, Mark, & Regan, Lynne. (2000). Proteomics of *Mycoplasma genitalium*: Identification and characterization of unannotated and atypical proteins in a small model genome. *Nucleic Acids Research*, **28**, 3075–82.

Baltimore, D. (1970). RNA-dependent DNA polymerase in virons of RNA tumour viruses. *Nature*, **226**, 1209–11.

Baly, E.C.C. (1928). Photosynthesis. *Science*, **LXVIII**, 364–7.

Baly, Edward Charles Cyril, Heilbron, Isidor Morris, & Hudson, Donald Pyrice. (1922). CXXX–Photocatalysis. Part II, The Photosynthesis of Nitrogen Compounds from Nitrates and Carbon Dioxide. *Journal of the Chemical Society*, **121**, 1078–88.

Bandfield, Joshua L. Glotch, Timothy D., & Christensen, Philip R. (2003). Spectroscopic identification of carbonate minerals in the Martian dust. *Science*, **301**, 1084–7.

Banin, A., & Navrot, J. (1975). Origin of life: Clues from relations between chemical composition of living organisms and natural environments. *Science*, **189**, 550–1.

Barber, David, J., & Scott, Edward, R. D. (2002). Origin of supposedly biogenic magnetite in the Martian meteorite Allan Hills 84001, *Proceedings National Academy of Sciences*, **99**, 6556–61.

Barrell, B. G., Air, G. M., & Huchinson, C. A. III. (1976). Overlapping genes in bacteriophage ΦX174. *Nature*, **264**, 34–41.

Barrell, B. G., Anderson, S., Bankier, A. T., de Bruijn, M. H. L., Chen, E., Coulson, A. R., Drouin, K., Eperon, I. C., Nierlich, D. P., Roe, B. A., Sanger, F., Schreirer, P. H., Smith, A. J. H., Stadem, R., & Young, I. G. (1980). Different pattern of codon recognition by mammalian mitochrodrial tRNAs. *Proceedings of the National Academy of Sciences,* **77**, 3164–6.

Barrell, B. G., Bankier, A. T., & Drouin, J. (1979). A different genetic code in human mitochondria. *Nature*, **282**, 189–94.

Battail, Gérard. (2001). Is biological evolution relevant to information theory and coding? *Proc. ISCTA 2001*, 343–51 Ambleside, UK.

Baudisch, Oskar. (1913). Über Nitrat-und Nitritassimilation *Zeitschrift der Angewandte Chemie*, **26**, 612–13.

Begley, Sharon, & Rogers, Adam. (1997). War of the worlds. *Newsweek*, February 10, 1997.

Behe, Michael J. (1996). *Darwin's Black Box: The Biochemical Challenge to Evolution*. New York: Free Press, Simon & Schuster.

Behe, Michael J., Dembski, William A., & Meyer, Stephen C. (2002). *Science and Evidence for Design in the Universe*. San Francisco: Ignatius Press.

Bellman, Richard. (1997). *Introduction to Matrix Analysis*. Philadelphia: Society for Industrial and Applied Mathematics.

Bennett, C. H. (1973). Logical reversibility of computation. *IBM Journal of Research and Development*, **17**, 525–32.

Bennett, C. H. (1988). Notes on the history of reversible computation. *IBM Journal of Research and Development*, **32**, 16–23.

Bennett, C. H., & Landauer, R. (1985). The fundamental physical limit of computation. *Scientific American*, **253**, 48–56.

Bergson, Henri-Louis. (1944). *Creative Evolution*. Authorized translation by Arthur Mitchell. New York: Modern Library.

Berman, Abraham, & Plemmons, Robert J. (1994). *Nonnegative Matrices in the Mathematic Sciences*. Philadelphia: Society for Industrial and Applied Mathematics.

Bernal, J. D. (1951). *The Physical Basis of Life*. London: Routledge and Paul.

Bernal, J. D. (1967). *The Origin of Life*. London: Weidenfeld & Nicolson.

Bernstein, Max P., Dworkin, Jason P. Sandiford, Scott A., Cooper, George W., & Allamandolla, Louis J. (2002). Racemic amino acids from the ultraviolet proteolysis of interstellar ice analogues. *Nature*, **416**, 401–3.

Berry, M. J., Banu, L., & Larsen, P. R. (1991). Type I iodothyronine deiodinase is a selenocysteine-containing enzyme. *Nature*, **349**, 438–40.

Bertram, Gwyneth, Innes, Shona, Minella, Odile, Richardson, Jonathan P., & Stansfield, Ian (2001). Endless possibilities: Translation termination and stop codon recognition. *Microbiology*, **147**, 255–69.

Berzelius, J. J., & Wöhler, F. (1901). *Briefwechsel zwischen Berzelius und. Wöhler Berzelius und Liebig ihre Briefe von 1831–1845 mit ertäuternden Einschaltuncen aus gleichzeitigen Briefen von Liebig und Wöhler sowie Wissenschaftlichen nachweisen microform: herausgegben mit unterstützung der Kgt. Bayer. Akademie der Wissenschaften von Justus Carrière.*

Bibb, M. J., van Etten, R. A., Wright, C. T., Walberg, M. W., & Clayton, D. A. (1981). Sequence and gene organization of mouse mitochondrian DNA. *Cell*, **26**, 167–80.

Billingsley, P. (1965). Ergodic Theory and Information. (see Theorem 15, in Chapter 5) New York, London, Sydney: John Wiley.

Billingsley, Patrick. (1995). *Probability and Measure*, third edition. New York: John Wiley.

Bizzarro, Martin, Bajer, Joel A. Haack, Henning, Ulfbeck, David, & Rosing, Minik. (2003). Early history of Earth's crust-mantle system inferred from hafnium isotopes in chondrites. *Nature*, **421**, 931–3.

Böck, August. (2002). Invading the genetic code. *Science*, **292**, 453–4.

Bohr, N. (1933). Light and Life. *Nature*, **308**, 421–3; 456–9.

Bold, Benjamin. (1982). *Famous Problems in Geometry and How to Solve Them.* New York: Dover Publications.

Bongaarts, John, & Feeney, Griffith. (2003). Estimating mean lifetime. *Proceedings of the National Academy of Sciences*, **100**, 13127–33.

Bonitz, S., Berlani, R., Coruzzi, G., Li, M., Macino, G., Nobrega, F. G., Nobrega, M. P., Thalenfeld, B. E., & Tzagoloff, A. (1980). Codon recognition rules in yeast mitochondria. *Proceedings of the National Academy of Sciences*, **77**, 3167–70.

Borstnik, P., & Hofacker, G. I. (1985). Functional aspects of the neutral patterns in protein evolution. In *Structure & Motion, Nucleic Acids & Proteins*, eds. E. Clementi, G. Corongiu, M. H. Sarma & R. H. Sarma. New York: Academic Press.

Borstnik, P., Pumpernik, D., & Hofacker, G. I. (1987). Point mutations as an optimal search process in biological evolution. *Journal of Theoretical Biology*, **125**, 249–68.

Bové, J. M. (1984). Wall-less prokyrotes of plants. *Annual Reviews of Phytopathology*, **22**, 361–96.

Bowers, John E., Chapman, Brad A., Rong, Junkang, & Paterson, Andrew H. (2003). Unraveling angiosperm genome evolution by phylogenetic analysis of chromosomal duplication events. *Nature*. **422**, 428–433.

Bradley, J. P., Harvey, H. Y., & McSween Jr. (1997). No nanofossils in Martian meterorite. *Nature*, **390**, 454–6.

Brantly, M., Courtney, M., & Crystal, R. G. (1988). Repair of the secretion defect in the Z form of α1-antitrypsin by addition of a second mutation. *Science*, **242**, 1700–1.

Brasier, Martin, D., Green, Owen R., Jephcoat, Andrew, Kleppke, Annette K., Van Kranendonk, Martin J., Lindsay, John F., Steele, Andrew, & Grassineau, Nathalie. (2002). *Science*, **416**, 76–81.

Breiman, L. (1957/1960). The individual ergodic theorem of information theory. *Ann. Math. Stat.*, **28**, 809–11; Correction, *Ibid.*, **31**, 890–10.

Brillouin, L. (1953). The negentropy principle of information. *Journal of Applied Physics*, **24**, 1153.

Brillouin, L. (1962). *Science and Information Theory.* second edition. New York: Academic Press.

Brillouin, L. (1990). Life, thermodynamics, and cybernetics. In *Maxwell's Demon, Entropy, Information, Computing.* Princeton, NJ: Princeton University Press.

Bungenburg de Jong, H. G. (1932). Die Koazervation und ihre Bedeutung für die Biologie. *Protoplasma*, **15**, 110–73.

Burch, Douglas. (2003). Key seed bank may be uprooted. *The Sun*, April 27, p. 2A.

Burke, Stephen, Lo, Sam L., Krzycki, Joseph. (1998). Clustered genes encoding the methyltranfererase of methanogenesis from monomethylamine. *Journal of Bacteriology*, **180**, 3432–40.

Buseck, Peter R., Dunin-Borkowski, Rafal E., Devouard, Bertrand, Frankel, Richard, Richard B. Mccartney, Martha R., Midgley, Paul A., Pósfai, Mijály, & Weyland, Matthew. (2001). Magnetic Morphology and Life on Mars. *Proceedings of the National Academy of Sciences*, **98**, 13590–495.

Bushman, Frederic. (2002). *Lateral DNA Transfer Mechanisms and Consequences*. New York: Cold Spring Harbor Laboratory Press.

Butler, Declan. (2004). Mars satellite flies into hunt for lost Beagle 2. *Nature*, **427**, 5.

Byerly, Gary R., Lowe, Donald R., Wooden, Joseph L., & Xiaogang Xie. (2002). An Archean Impact Layer from the Pilbara and Kaapvaal Cratons. *Science*, **297**, 1325–7.

Cairns-Smith, A. G. (1965). The origin of life and the nature of the primitive gene. *Journal of Theoretical Biology*, **10**, 53–88.

Cairns-Smith, A. G. (1971). *The Life Puzzle*. Edinburgh: Oliver & Boyd.

Cairns-Smith, A. G. (1982). *Genetic Takeover and the Mineral Origin of Life*. Cambridge, UK: Cambridge University Press.

Calude, C., & Chaitin, G. (1999). Mathematics: Randomness everywhere. *Nature*, **400**, 319–20.

Calvin, M. (1961). *Chemical Evolution*. Oxford: University Press.

Canup, R., & Asphaug, E. (2001). Origin of the Moon in a giant impact near the end of the Earth's formation. *Nature*, **412**, 708–12.

Canup, R. M., & Righter, K. (eds.). (2000). *The Origin of the Earth and the Moon*. Tucson: University of Arizona Press, in collaboration with the Lunar and Planetary Institute, Houston.

Canuto, V. M., Levine, J. S., Augustsson, T. R., Imhoff, C. L., & Giampapa. (1983). The young Sun and the atmosphere and photochemistry of the early Earth. *Nature*, **305**, 281–6.

Caro, G. M. Muñoz, Meierhenrich, U. J., Schjutte, W. A., Barbier, B., Segovia, A. Arcones, Rosenbauer, H., Theimann, W. H.-O., Brack, A., & Greenberg, J. M. (2002). Amino acids from ultraviolet irradiation of interstellar ice analogues. *Nature*, **416**, 403–6.

Caro, Guillaume, Bourdon, Bernard, Birck, Jean-Louis & Moorbath, Stephen. (2003). [146]Sm[143]Nd evidence from Isua metamorphosed sediments for early differentiation of the Earth's mantle. *Nature*, **432**, 428–32.

Castresana, Jose, Feldmair-Fuchs, Gertraud, & Pääbo, Svante. (1998). Codon reassignment and amino acid composition in hemichordate mitochondria, *Proceedings of the National Academy of Sciences*, **98**, 3703–7.

Cayrel, R., Hill, T. C., Beers, B., Barbuy, M., Spite, F., Spite, B., Pelz, J., Andersen, P., Bonifacio, P., François, P., Molaro, B., Nordström, F. Primas. (2001). Measurement of stellar age from uranium decay. *Nature*, **409**, 691–2.

Cech, T. R. (1986). A model for the RNA-catalyzed replication of RNA. *Proceedings National Academy of Sciences*, **83**, 4360–3.

Chaisson, Eric J. (2001b). *Cosmic Evolution: The Rise of Complexity in Nature*. Cambridge, MA: Harvard University Press.

Chaitin, G. (1990). *Information, Randomness, and Incompleteness*. Singapore: World Scientific.

Chaitin G. (1992a). *Information-Theoretic Incompleteness*. Singapore: World Scientific.

Chaitin, G. (1992b). Randomness in arithmetic and the decline and fall of reductionism in pure mathematics. *IBM Research Report* RC-18532.

Chaitin, G. (1999). *The Unknowable*. Singapore: Springer Verlag.

Chaitin, G. J. (1966). On the length of programs for computing finite binary sequences. *Journal of the Association for Computing Machinery*, **13**, 547–69.

Chaitin, G. J. (1975). A theory of program size formally identical to information theory. *Journal of the Association for Computing Machinery*, **22**, 329–40.

Chaitin, G. J. (1985). An APL2 gallery of mathematical physics–a course outline. *Proceedings Japan 85 APL Symposium Publication N:GE18-9948-0, IBM Japan*, 1–56.

Chaitin, G. J. (1987a). *Algorthmic Information Theory*. Cambridge, UK: Cambridge University Press.

Chaitin, G. J. (1987b). Incompleteness theorems for random reals. *Advances in Applied Mathematics*, **8**, 119–46.

Chaitin, Gregory, J. (1979). Toward a mathematical definition of life. In R. D. Levine & M. Tribus, eds., *The maximum Entropy formalism*. Cambridge, MA and London MIT Press.

Chaitin, Gregory J. (2001a). *Exploring Randomness*. Singapore: Springer Verlag.

Chaitin, Gregory J. (2001b). *The Limits of Mathematics–A Course on Information Theory and the Limits of Formal Reasoning*. New York: Springer-Verlag.

Chambers, I., Frampton, J., Goldfarb, P., McBain, W., & Harrison, P. R. (1986). The structure of the mouse glutathione peroxidase gene: The selenocysteine in the active site is encodded by the "termination" codon, TGA, *The EMBO Journal*, **5**, 1221–7.

Christensen, Philip R., Bandfield, Joshua L., Bek III, James F., Gorelick, Noel, Hamilton, Victoria, E., Ivanov, Anton, Jakosky, Bruce M., Kieffer, Hugh H., Lane, Melissa D., Malin, Michael C., McConnonchie, Timothy, McEwen Alfred S., McSween, Jr., Harry, Y., Mehall, Greg L., Moersch, Jeffrey E., Nealson, Kenneth H., Rice, James W., Jr., Richardson, Mark I., Ruff, Steven W., Smith, Michael D., Titus, Timothy N., & Wyatt, Michael B. (2003). Morphology and composition of the surface of Mars: Mars odyssey THEMIS results. *Science*, **300**, 2056–61.

Chyba, Christopher, & Phillips, Cynthia B. (2002). Europa as an abode of life. *Origins of Life and Evolution of the Biosphere*, **32**, 47–68.

Clark, Andrew G., Glanowski, Stephen, Nielsen, Rasmus, Thomas Paul D., Kejariwal, Anish, Todd, Melissa A., Tanenbaum, David M., Civello, Daniel, Lu, Fu, Murphy, Brian, Ferriera, Steve, Wang, Gary, Zhengh, Xianqgun, White, Thomas J., Sninsky, John J., Adams, Mark D, & Cargil, Michele. (2003). Inferring nonneutral evolution from human-chimp-mouse orthologous gene trios. *Science*, **302**, 1960–3.

Clary, D. D., & Wolstenholme, D. R. (1985). The mitochondrial DNA molecule of *Drosophila yukuba*: Nucleotide sequence gene organization and genetic code, *Journal of Molecular Evolution*, **22**, 252–71.

Cohen, B. A, Swindle, T. D., & Kring, D. A. (2000). Support for the lunar cataclysm hypothesis from lunar meteorite impact melt ages. *Science*, **290**, 1754–6.

Cohen, Daniel. (2000). Genetic and physiological data implicating the new human gene G72 and the gene for D-amino acid oxidase in schizophrenia, *Proceedings of the National Academy of Sciences*, **99**, 13675–80.

Cohn, C. A., Hasson, T. K., Larsson, H. S. Sowerby, S, J., & Holm, N. G. (2001). Fate of prebiotic adenine. *Astrobiology*, **1**, 477–80.

Collie, J. N. (1905). Synthesis by means of the silent electrical discharge. *Journal of the Chemical Society*, **79**, 1540–8.

Collie, J. Norman (1901). On the decomposition of carbon dioxide when submitted to electric discharge at low pressures. *Journal of the Chemical Society*, **79**, 1063–9.

Commoner, Barry. (1964). Roles of desoxyribosenucleic acid in inheritance. *Nature*, **202**, 960–8.

Commoner, Barry. (1968). Failure of the Watson–Crick theory as a chemical explanation of inheritance. *Nature*, **220**, 334–40.

Commoner, Barry. (2002). The spurious foundation of genetic engineering. *Harpers Magazine*, February.

Cooper, George, Kimmich, Novelle, Belisle, Warren, Sarinana, Josh, Brabham, Katrina, & Garrel, Laurence. (2001). Carbonaceous meteroites as a source of sugar-related organic compounds for the early Earth. *Nature*, **414**, 879–83.

Cooper, P. R., Smilinich, N. J., Day, C. D., Nowak, N. J., Reid, L. H., Pearsall, R. S., Reece, M., Prawitt, Landers J., Housman, D. E., Winterpacht, A., Zabel, B. U., Pelletier, J., Weissman, B. E., Shows, T. B., & Higgins, M. J. (1998). Divergently transcribed overlapping genes in liver and kidney and located in the 11p15.5 imprinted domain. *Genomics*, **49**, 38–51.

Correia, Alexandre C., & Laskar, Jacques. (2001). The four final rotation states of Venus. *Nature*, **411**, 767–70.

Cosmochemistry & the Origin of Life: NATO Advanced Study Institutes Series, *NATO Advanced Study Institute, Cyril Ponnampeuma, North Atlantic Treaty Organization Scientific Affairs Division*. New York: Kluwer Academic Publishers (April 1983).

Crick, F. H. C., Griffith, J. S., & Orgel, L. E. (1957). Codes without commas. *Proceedings of the National Academy of Sciences*, **43**, 416–21.

Crick, F. H. C. (1968). The origin of the genetic code, *Journal of Molecular Biology*, **38**, 367–79.

Crick, Francis. (1970). Central dogma of molecular biology. *Nature*, **227**, 561–3.

Crick, Francis. (1981). *Life Itself, Its Origin, and Nature*. New York: Simon & Schuster.

Cronin, John R., & Pizzarello, Sandra. (1997). Enantiomeric excess in meteoritic amino acids. *Science*, **275**, 951–6.

Croty, Shane, Cameron, Craig E., & Andino, Raul. (2001). RNA virus error catastrophe: Direct molecular test by using ribavirin, *Proceedings of the National Academy of Sciences*, **98**, 6895–900.

Culler, Timothy S., Becker, Timothy A., Muller, Richard A., & Renne, Paul R. (2000). Lunar history from $^{40}Ar/^{39}$ dating of glass spherules. *Science*, **287**, 1785–8.

Cullmann, G. (1981). A mathematical method for the enumeration of doublet codes. In *Origin of Life*, ed. Y. Wolman (pp. 405–13). Dordrecht: Reidel.

Cullmann, G., & Labougues. J.-M. (1987). Evolution of proteins: An ergodic stationary chain. *Mathematical Modeling*, **8**, 635–46.

Cullmann, G., & Labouygues, J.-M. (1983). Noise immunity in the genetic code. *BioSystems*, **16**, 9–29.

Cupples, C. G., & Miller, J. H. (1988). Effects of amino acid substitutions at the active site in Escherichia coli β-galactosidase. *Genetics*, **120**, 637–44.

Dalton, Rex. (2002). Microfossils: Squaring up over ancient life. *Nature*, **417**, 782–4.

Darwin, Charles Robert. (1872). *The Origin of Species by Means of Natural Selection or the Preservation of Favored Races in the Struggle for Life*. New York and Scarborough: Ontario.

Darwin, F. (1898). *The Life and Letters of Charles Darwin*, **II**. New York: D. Appleton.

Das, G., Hickey, D. R., McLendon, D., McLendon, G., & Sherman, F. (1989). Dramatic thermostabililzation of yeast iso-1-cytochrome c by an asparagine-isolucine replacement at site 57, *Proceedings of the National Academy of Sciences*, **83**, 1271–5.

Davis, Mark A. (2003). A History Lesson for President Putin? *Science*, **300**, 249.

Davis, Paul. (1999). *The Fifth Miracle: The Search for the Origin and Meaning of Life.* New York: Simon & Schuster.

Davis, Wanda L., & McKay, Christopher P. (1966). Urey Prize Lecture: *Origins of Life and Evolution of the Biosphere*, **26**, 61–73.

Dawkins, Richard. (1995). *River out of Eden: A Darwinian view of Life.* New York: Basic Books, A Division of HarperCollins Publishers.

Dawkins, Richard. (1996). *Climbing Mount Improbable.* New York: W. Norton & Company.

Dayhoff, M. (1976). *Atlas of Protein Sequences and Structure*: Vol. 5, *Supplement 2.* Silver Spring, MD: National Biomedical Research Foundation.

Dayhoff, M., & Eck, R. V. (1978). *Atlas of Protein Sequences and Structure*: Vol. 5, *Supplement 3*, National Biomedical Research Foundation.

Dayhoff, M. O., Eck, R. V., & Park, C. M. (1972). A model of evolutionary change in proteins. In *Atlas of Protein Sequence and Structure*, Vol. 5, M. O. Dayhoff, ed. (pp. 89–100). Silver Spring, MD: National Biomedical Research Foundation.

de Duve, Christian. (1991). *Blue Print for a Cell.* Burlington, NC: Patterson Publishers, Carolina Biological Supply company.

de Duve, Christian (1995). *Vital Dust: Life as A Cosmic Imperative.* New York: Basic Books.

Deamer, D. (1997). The first living systems: A bioenergic perspective. *Microbiology and Molecular Biology Reviews*, **61**, 239–61.

Deamer, David W., & Fleischaker, Gail F. (1994). *Origins of Life: The Central Concepts.* Boston: Jones & Bartlett.

Dembski, William A. (1998a). *The Design Inference, Eliminating Chance through Small Probabilities.* Cambridge, UK: Cambridge University Press.

Dembski, William A. (1998b). The intelligent design movement. *Cosmic Pursuit*, **1**, 22–6.

Dembski, William A. (1999). *Intelligent Design: The Bridge between Science and Theology.* Downers Grove, IL: InterVarsity Press.

Dembski, William A. (2002). *No Free Lunch: Why Specified Complexity Cannot be Purchased without Intelligence.* Lanham, MD: Rowman & Littlefield.

Dembski, William A., & Ruse, Michael. (2004). *Debating Design.* Cambridge, UK: Cambridge University Press.

Diaconis, P. W., & Holmes, S. P. (1998). Matchings and phylogenetic trees, *Proceedings of the National Academy of Sciences*, **95**, 14600–2.

Doolittle, R. F. (1981). Similar amino acid sequences: Chance or common ancestry? *Science*, **214**, 149–59.

Doolittle, R. F. (1987a). The evolution of the vertebrate plasma proteins. *Biological Bulletin*, **172**, 269–83.

Doolittle, R. F. (1987b). *Of URFS and OGFS: A primer on how to analyze derived amino acid sequences.* Mill Valley, CA: University Science Books.

Doolittle, R. F. (1988). More molecular opportunism. *Nature*, **336**, 18.

Doolittle, W. Ford (1999). Phylogenetic classification and the universal tree. *Science*, **284**, 2124–8.

Doolittle, W. Ford. (2000). The nature of the universal ancestor and the evolution of the proteome. *Current Opinion in Structural Biology*, **10**, 355–8.

Doudna, Jennifer A., & Cech, Thomas R. (2000). The chemical repertoire of natural ribosomes. *Nature*, **418**, 222–8.

Doyle, John. (2001). Computational biology: Beyond the spherical cow. *Nature*, **411**, 151–2.

Driesch, Hans. (1914). *The History and Theory of Vitalism*, authorized translation by C. K Ogden. London: Macmillan.

Durham, A. (1978). *New Scientist*, **77**, 785–7.

Dyson, F. J. (1982). A model for the origin of life. *Journal of Molecular Evolution*, **18**, 344–50.

Edelman, Gerald M., & Gally, Joseph M. (2001). Degeneracy and complexity in biological systems, *Proceedings of the National Academy of Sciences*, **98**, 13763–8.

Edwards, A. W. F., & Cavalli-Sforza, L. L. (1964). In *Phenetic and Phylogenetic Classification*, V. H. Heywood and J. McNeil, eds. (pp. 67–76) London: Symatics Association, Publication No. 6.

Eigen, M. (1971). Self-organization of matter and the evolution of biological macromolecules. *Naturwissenschaften*, **58**, 465–523.

Eigen, Manfred. (1992). *Steps toward Life*. Oxford: Oxford University Press.

Eigen, Manfred. (2002). Error catastrophe and antiviral strategy, *Proceedings of the National Academy of Sciences*, **99**, 13374–6.

Eigen, Manfred. (1993). The origin of genetic information: Viruses as a model. *Gene*, **135**, 37–47.

Eigen, Manfred. (1977). The hypercycle: A principle of natural self-organization. Part A: Emergence of the hypercycle. *Naturwissenschaften*, **64**, 541–65.

Eigen, Manfred. (1978a). The hypercycle: A principle of natural self-organization. Part B: The abstract hypercycle. *Naturwissenschaften*, **65**, 7–41.

Eigen, Manfred. (1978b). The hypercycle: A principle of natural self-organization. Part C: The abstract hypercycle. *Naturwissenschaften*, **65**, 341–69.

Eigen, Manfred, & Schuster, Peter. (1979). *The Hypercycle: A Principle of Natural Self Organization*. Berlin: Springer Verlag.

Eigen, Manfred, Schuster, Peter. (1982). Stages of emerging life–five principles of early organization, *Journal of Molecular Evolution*, **19**, 47–61.

Eigen, Manfred, Winkler-Oswatitsch, Ruthild, & Dress, Andreas. (1988). Statistical geometry in sequence space: A method of quantitative comparative sequence analysis. *Proceedings of the National Academy of Sciences*, **85**, 5913–17.

Einstein, A. (1905). Über die von molecular-kinetischen Theorie der Wärme geforderte Bewegung in ruhenden flüssigkeiten suspendierten Teilchen. *Annalen der Physik*, **17**, 549–60.

Einstein, A. (1906). Zur Theorie der Brownischen Bewegung. *Annalen der Physik*, **19**, 371–81.

Elkin, Lynne Osman. (2003). Rosalind Franklin and the double helix. *Physics Today*, **56**, 42–8.

Emiliani, Cesare. *Planet Earth*. Cambridge, UK: Cambridge University Press, p. 372.

Engels, F. (English translation 1954). *The Dialectics of Nature*. Moscow: Foreign Language Publication House.

Essene, E. J., & Fisher, D. C. (1986). Lightning strike fusion: extreme reduction and metal-silicate liquid immiscibility, *Science*. **234**, 189–93.

Eves, Howard. (1966). *Elementatry Matrix Theory*. New York: Dover Publications, Inc.

Farina, M., Esquivel, D. M., & de Barros, H. G. P. L. (1990). Magnetic iron-sulfur crystals from a magnetotactic microorganism. *Nature*, **343**, 256–8.

Feller, W. (1968). *An Introduction to Probability Theory and its Applications*. Third edition. New York: Wiley & Sons.

Ferris, James P. Hill, Aubrey R., Liu, Rihe, and Orgel, Leslie E. (1996). Synthesis of long prebiological oligomers on mineral surfaces. *Nature*, **381**, 59–61.

Fiddes, J. C. (1977). The nucleotide sequences of a viral DNA. *Scientific American*, **237**, 55–67.

Fiers, W., Contreras, R., Haegman, G., Rogiers, R., van der Voorde, A., van Heuverswyn, H., van Herreweghe, J., Volckaert, G., & Ysebaert, M. (1978). Complete sequence of SA 40 DNA. *Nature*, **273**, 113–20.

Figureau, A., & Labouygues, J.-M. (1981). The origin and evolution of the genetic code. In *Origin of Life*, Y. Wolman, ed. Dordrect: Reidel.

Fisher, Ronald Aylmer. (1930). *The Genetical Theory of Natural Selection*. Oxford: Oxford University Press.

Fox, S. W. et al. (1994). Experimental retracement of the origins of a protocell, it was also a protoneuron. *Journal of Biological Physics*, **20**, 17–36.

Fox, S. W. et al. (1996). "Experimental retracement of terrestrial origin of an excitable cell: *Was it Predictable?*" In *Chemical Evolution*, J. Chela-Flores & F. Raulin, eds. The Netherlands: Kluwer Academic Publishers.

Fox, T. D. (1987). Natural variations in the genetic code. In *Annual Reviews of Genetics*, A. Campbell, I. Herkowitz, & L. M. Sander, eds. pp. 67–91.

Freedman, M. H. (1998). Limit, logic and computation. *Proceedings of the National Academy of Sciences*, **95**, 95–7.

Freedman, Wendy L., & Feng. Long Long. (1999). The determination of the Hubble constant *Proceedings of the National Academy of Sciences*, **96**, 11063–64.

Freeland, Stephen J., Knight, Robin D., & Landweber, Laura F. (1999). Do Proteins predate DNA? *Science*, **286**, 690–2.

Freeland, Stephen J., Knight, Robin D., Lansweber, Laura F., & Hurst, Laurence D. (2000). Eary fixation of an optimal genetic code. *Molecular Biology and Evolution*, **17**, 511–18.

Freist, W. et al. (1998). Accuracy of protein biosynthesis: Quasi-species nature of protein and possibility of error catastrophes, *Journal of Theoretical Biology*, **193**, 19–38.

Freistropher, David V., Kwiatkowski, Marek, Buckingham, Richard H., & Ehrenberg, Måns. (2000). The accuracy of codon recognition by polypeptide release factors, *Proceedings of the National Academy of Sciences*, **97**, 2046–51.

Frobenius, G. (1908). Über Matrizen aus positiven Elementen. *Sitzungsberichte der Königliche Preussische Akademie der Wisschenschaft*, 514–18.

Frobenius, G. (1912). Über Matrizen aus nicht negativen Elementen. *Sitzungsberichte der Königliche Preussische Akademie der Wisschenschaft*, 456–77.

Füchslin, Rudolf M., & McCaskill, John S. (2001). Evolutionary self-organization of cell-free genetic coding. *Proceedings of the National Academy of Sciences*, **98**, 9185–90.

Fukuda, Yoko, Washio, Takanori, & Tomita, Masaru. (1999). Comparative study of overlapping genes in the genomes of *Mycoplasma genitalium* and *Mycoplasma pneumoniae*. *Nucleic Acids Research*, 1847–53.

Gamow, George, & Ycas, Martynus. (1955). Statistical correlation of protein and ribonucleic acid composition. *Proceedings of the National Academy of Sciences*, **41**, 1011–19.

Gamow, George. (1954a). Possible relation between Deoxyribonucleic Acid and Protein Structures. *Nature*, **173**, 318.

Gamow, George. (1954b). Possible mathematical relation between deoxyribonucleic acid and proteins. *Det Kong. Danske Vid. Selskab*, **22**, 1–13.

Gamow, George. (1961). What is life? *Trans. Bose Rea. Inst.*, **24**, 185–92.

García-Ruiz, J. M., Hyde, S. t. Carnerup A. M. Christy, A. G. Van Kranendonk, M. J., & Welham, N. J. (2003). Self-assembled silica-carbonate structures and detection of ancient microfossils. *Science*, **302**, 1194–7.

Gardell, S. J., Craik, C. S., Hilvert, D., Urdea, M. S., & Rutter, W. J. (1985). Site-directed mutagenesis shows that tyrosine 248 of carboxypeptide A does not play a crucial role in catalysis. *Nature*, **317**, 551–5.

Geller, A. I., & Rich, A. (1980). A UGA termination suppression of tRNA[Trp] active in rabbit reticulocytes. *Nature*, **283**, 41–6.

George, D. G., Hunt, L. T., Yeh, L.-S., & Barker, W. C. (1985). New perspective on bacterial ferrodoxin evolution. *Journal of Molecular Evolution*, **22**, 20–32.

Gil, Rosario, Sabater-Muñoz, Betriz, Latorre, Amparo, Silva, Francesco J., & Moya, Andrés. (2002). Extreme genome reduction in Buchnera ssp.: Toward the minimal genome needed for symbiotic life, *Proceedings of the National Academy of Sciences*, **99**, 4454–8.

Gilbert, W. (1986). The Origin of life: The RNA world. *Nature*, **319**, 618.

Gilbert, Walter. (1987). The exon theory of genes. *Cold Spring Harbor Symp. Quant. Biol.*, **52**, 901–5.

Gilbert, Walter. (2003). Life after the helix. *Nature*, **421**, 315.

Gilbert, Walter, de Souza, Sandro J., & Long, Manyuan. (1997). "Origin of genes" *Proceedings of the National Academy of Sciences*, **94**, 7698–703.

Gilbert, Walter, de Souza, & Sandro J. (1999). *Introns and the RNA World*. In *The RNA World*, second Edition, Cold Spring Harbor Laboratory Press.

Glavin, Daniel P. Bada, Jeffrey L., Brinton, L. F., & McDonald, Gene D. (1999). Amino acids in the Martian meteorite Nakhla, *Proceedings of the National Academy of Sciences*, **96**, 8835–8.

Glockler, G., & Lind S. C. (1939). *The Electrochemistry of Gases and Other Dielectrics*. New York: John Wiley & Sons.

Gödel, Kurt. (1931). Über formal unentscheidbare Sätze der Principia Mathematica und verwandter Systeme I. *Monatshefte für Mathematik und Physik*, **38**, 174–98.

Gödel, Kurt. (1992). *On Formally Undecidable Propositions of Principia Mathematica and Related Systems*. Trans. B. Meltzer, with Intro. by R. B Brathwaite. New York: Dover.

Godfrey-Smith, Peter. (2000a). On the Theoretical Role of Genetic Coding. *Philosophy of Science*, **67**, 26–44.

Godfrey-Smith, Peter. (2000b). Information, Arbitrariness, and Selection: Comments on Maynard Smith. *Philosophy of Science*, **67**, 202–7.

Gompertz, B. (1825). On the nature of the function expressive of the law of human mortality and on a new mode determining life contingencies, *Philosophical Transaction of the Royal Society of London*, **II**, 513–85.

Gough, D. O. (1981). Solar interior structure and luminosity variations. *Solar Physics*, **74**, 21–34.

Graham, L. R. (1993). *Science in Russia and the Soviet Union*. Cambridge, UK: Cambridge University Press.

Graham, Loren R. (1972). *Science and Philosophy in the Soviet Union*. New York: Alfred A. Knopf.

Graham, Loren R. (1987). *Science, Philosophy and Human Behavior in the Soviet Union*. New York: Columbia University Press.

Grande-Pérez, Sierra, S., Castro, M. G., Domingo, E., & Lowenstein, P. R. (2002). Molecular determination in the transition to error catastrophe: Systematic elimination of lymphocytic choriomeningitis virus through mutagenesis does not correlate linearly with large increases in mutant spectrum complexity, *Proceedings of the National Academy of Sciences*, **99**, 12938–43.

Graure, D., & Li, W. H. (2000). *Fundamentals of Molecular Evolution*. Sunderland, MA: Sinauer.

Griffith, J. S. (1967). Self-replication and scrapie. *Nature*, **215**, 1043–4.

Gu, Zhenlong, Steinmetz, Lars M., Gu, Xun, Scharfe, Curt, Davis, Ronald W., & Li, Wen-Hsiung. (2003). Role of duplicate genes in genetic robustness against null mutations. *Nature*, **421**, 63–6.

Haeckel, E. H. P. A. (1866). Entstehung der ersten Organismen. In *Generelle Morphologie der Organismen allgemeine Grundzüge der organischen Formen-Wissenchaft: Mechanisch begründet durch die von Charles Darwin reformirte Descendenz-Theorie* VI. Berlin: George Reimer.

Haeckel, Ernst. (1905). *The Wonders of Life*. New York: Harper.

Hagmann, Michael. (2002). Between a rock and a hard place. *Science*, **295**, 2006–7.

Haldane, J.B.S. (1929). The Origin of Life. *The Rationalist Annual*. London: 242–9.

Haldane, J.B.S. (1932). *Causes of Evolution*. London: Longmans and Green.

Haldane, John B. S. (1954). The origins of life. *New Biology*, **16**, 12–27.

Haldane, John Burdon Sanderson. (1927). *Possible Worlds and Other Essays*. London: Chatto & Windus.

Halliday, Alex N. (2004). Mixing, volatile loss and compositional change during impact-driven accretion of the Earth. *Nature*, **427**, 505–9.

Hamming, R.W. (1950). Error detecting and error correcting codes. *Bell System Technical Journal*, **29**, 147–60.

Hamming, R. W. (1986). *Coding and Information Theory*. Englewood Cliffs, NJ: Prentice Hall.

Hao Bing, Gong, Weimin, Ferguson, Tsuneo, K. James, Carey M., Krzycki. Joseph A., & Chan, Michael. (2002). A new UAG-encoded residue in the structure of a methanogen methyltransferase. *Science*, **296**, 1462–6.

Hardin, Garrett. (1950). Darwin and the Heterotroph Hypothesis. *Scientific Monthly*, 178–9.

Harris, Joel Chandler. (1983). *The Complete Tales of Uncle Remus*. New York: Houghton-Mifflin Company.

Hart, M. H. (1978). The evolution of the atmosphere of the Earth. *Icarus*, **33**, 23–7.

Harvati, Katerina, Frost, Stephen R., & McNulty, Kieran P. (2004). Neanderthal taxonomy reconsidered: Implication of 3D primate models of intra-and interspecific differences. *Proceedings of the National Academy of Sciences*, **101**, 1147–52.

Hawks, W. C., & Tappel, A. L. (1983). In vitro synthesis of gluathione peroxidase from selenite translational incorporation of selenocystene. *Biochemica et Biophysica Acta*, **793**, 225–34.

Hazen, Robert M., Filley, Timothy R., & Goodfriend, Glenn A. (2001). Selective adsorption of L- and D-amino acids on calcite: Implication for biochemical homochirality. *Proceedings of the National Academy of Sciences*, **98**, 5487–90.

Head, James W., Mustard, John F., Kreslavsky, Mikhail A., Milliken, Ralph E., & Marchant, David R. (2003). Recent ice ages on Mars. *Nature*, **426**, 797–802.

Heckman, J. E., Sarnoff, J., Alzner-DeWeerd, B., Yin, S. & Raj Bandrary, U. L. (1980). Novel feature in the genetic code and codon reading patterns in *Neurospora crassa* mitochondria based on sequences of six mitochondrial tRNAs. *Proceedings of the National Academy of Sciences*, **77**, 3159–63.

Heilbron, J. L., & Bynam, W. F. (2002). 1902 and all that. *Nature*, **415**, 15–18.
Heilbron, J. L., & Seidel, Robert W. (1989). *Lawrence and His Laboratory: A History of the Lawrence Berkeley Laboratory, Volume 1*. California University Press.
Heisel, R., & Brennicke, A. (1983). Cyochrome oxidase subunit II gene in mitochondria of *Oenothera* has no intron. *The EMBO Journal* **2**, 2173–8.
Hekimi, Siegfried, & Guarente, Leonard. (2003). Genetics and the specificity of the aging process. *Science*, **299**, 1351–4.
Henikoff, J. E., Sarnoff, Keene, M. A., Fechtel, K., & Fristrom, J. (1986). Gene within a gene: Nested Drosphila genes encode unrelated proteins on opposite DNA strands. *Cell*, **44**, 33–42.
Henikoff, Stephen. (2002). Beyond the Central Dogma. *Bioinformatics*, **18**, 223–5.
Herken, Gregg. (2002). *Brotherhood of the Bomb*. New York: Henry Holt and Company.
Hermes, H., & Markwald, W. (1974). Foundations of mathematics. In *Fundamentals of Mathematics*, Vol. 1, H. Benke, F. Bachmann, K. Fladt, & W. Süss, eds. (pp. 1–80). Cambridge, MA: MIT Press.
Hilbert, David. (1900). Mathematische Probleme *Göttinger Nachrichten*. pp. 253–97. English translation by Dr. Mary Winston Newson. In *Bulletin of American Mathematical Society* (1902) **8**, 437–79.
Hoang, Linh, Bédarg, Sabrina, Krishna, Mallela M. G., Lin, Yan, & Englander, S. Walther. (2002). *Proceedings of the National Academy of Sciences*, **99**, 12173–8.
Hoefen, Todd M., Clark, Roger N. Bandfield, Smith, Michael, D., Pearl, John C., & Christensen, Philip R. (2003). Discovery of Olivine in the Nili Fossae region of Mars. *Science*, **302**, 627–30.
Hoffer, Eric. (1951). *The True Believer*. New York: Harper & Row.
Holliday, R. (1986). *Genes, Proteins, and Cellular Aging*. New York: Van Nostrand Reinhold.
Homer, (c. 850 B.C). *The Iliad*. a, Book III, lines 314–326; b, Book VII, lines 170–99; c, Book XV lines 158–217; d, Book III, lines 373–446 (a conversation between Aphrodite and Helen of Troy).
Hopfield, J. J. (1980). The energy relay: A proofreading scheme based on dynamic cooperativity and lacking all characteristic symptoms of kinetic proofreading in DNA replication and protein synthesis. *Proceedings of the National Academy of Sciences*, **77**, 5248–52.
Horowitz, N.H. (1986). *To Utopia and Back: The Search for Life in the Solar System*. New York: W. H. Freeman & Co.
Horowitz, N.H. (1990). Mission Impractical, *Science*, March–April, 44–9.
Horowitz, S., & Gorovsky, M. A. (1985). An unusual genetic code in nuclear genes of *Tetrahymena*, *Proceedings of the National Academy of Sciences*, **82**, 2452–5.
Hoyle, F., & Wickramasinghe, N. C. (1978). Life cloud: The Origin of Life in the Universe. New York: Harper & Row.
Huber, Claudia, & Wächtershäuser, Günter. (1997). Activated Acetic acid by carbon fixation on (Fe, Ni)S under primordial conditions. *Science*, **276**, 245–7.
Huber, Claudia, & Wächtershäuser, Günter. (1998). Peptides by activation of amino acids with CO on (Ni,Fe)S surfaces: Implications for the origin of life. *Science*, **281**, 670–2.
Huber, Claudia, Eisenreich, Wolfgang, Hecht, Stefan, & Wächterhäuser, Günter. (2003). A Possible primordial peptide cycle. *Science*, **301**, 938–40.

Hughes, Kimberly A., Alipaz, Julie A., Drnevich, Jenny, & Reynolds, Rose M., (2002). A test of evolutionary theories of aging. *Proceedings of the National Academy of Sciences*, **99**, 14286–91.

Hutchinson, Clyde A. III, Peterson, Scott N., Gill, Steven R. Cline, Robin T., White, Owen, Fraser, Claire M., Smith, Hamilton O., & Venter, J. Craig. (1999). Global Transposon Mutagenesis and a Minimal Mycoplasma Genome. *Science*, **286**, 2165–9.

Huynen, M. A., & Bork, P. (1998). Measuring genome evolution. *Proceedings of the National Academy of Sciences*, **95**, 5849–56.

Irion, Robert. (1998). Did twisty starlight set the stage for life? *Science*, **281**, 626–7.

James, Carey M., Ferguson, Tsunea K., Leykam, Joseph F., & Krzycki, Joseph A. (2001). The amber codon in the gene encoding the monomethylamine methyltransferase isolated from *Methanosarcina barkeri* is translate as a sense codon. *J. Biol. Chem.*, **276**, 34252–8.

Jankowski, J. M., Krawetz, S. A., Walcyzk, E., & Dixon, G. (1986). In vitro expression of two proteins from overlapping reading frames in a eukaryotic DNA sequence. *Journal of Molecular Evolution*, **24**, 61–71.

Jaynes, E. T. (1979). Where do we stand on maximum entropy? In *The Maximum Formalism*, R. D. Levine & M. Tribus, eds. (pp. 15–118). Cambridge, MA: MIT Press.

Jaynes, E. T. (1957a). Information theory and statistical mechanics. *Physical Review*, **106**, 620–30.

Jaynes, E. T. (1957b). Information theory and statistical mechanics, II. *Physical Review*, **108**, 171–90.

Jaynes, E. T. (1963). Information theory and statistical mechanics. In *Statistical Physics*, Vol. 3, G. E. Ulenbeck, N. Rosenzweig, A. J. F. Siegert, E. T, Jaynes, & S. Fujita, eds. New York: W. A. Benjamin.

Jenkin, H.C.F. (1867). The origin of species. *The North British Review*, **46**, 277–318.

Johnson, Philip E. (2000). *The Wedge of Truth: Splitting the Foundations of Truth.* New York: InterVarsity Press.

Joravsky, David. (1970). *The Lysenko Affair.* Cambridge, MA: Harvard University Press.

Joyce, G. F. (1998). Nucleic acid enzymes: Playing with a fuller deck. *Proceedings of the National Academy of Sciences*, **95**, 5845–47.

Joyce, G. F., et al. (1984). Chiral selection in poly(C)-directed synthesis of oligo(G), *Nature*, **310**, 602–4.

Joyce, Gerald F. (1991). The rise and fall of the RNA world. *The New Biologist*, **3**, 399–407.

Joyce, Gerald F. (2002a). The antiquity of RNA-based evolution. *Nature*, **418**, 214–21.

Joyce, Gerald F. (2002b). Molecular evolution: Booting up life. *Nature*, **420**, 278–9.

Jukes, T. H. (1965). Coding triplets and their possible evolutionary implications. *Biochemical and Biophysical Research Communications*, **19**, 391–6.

Jukes, T. H. (1966). *Molecules and Evolution.* New York: Columbia University Press.

Jukes, T. H. (1973). Possibilities for the evolution of the genetic code from a preceding form. *Nature*, **246**, 22–6.

Jukes, T. H. (1974). On the possible origin and evolution of the genetic code. *Origins of Life*, **5**, 331–50.

Jukes, T. H. (1980). Silent nucleotide substitution and the molecular evolutionary clock. *Science*, **210**, 973–8.

Jukes, T. H. (1981). Amino acid codes in mitochondria as possible clues to primitive codes. *Journal of Molecular Evolution*, **18**, 15–17.

Jukes, T. H. (1983a). Evolution of the amino acid code: Inference from mitochondrial codes. *Journal of Molecular Evolution*, **19**, 219–25.

Jukes, T. H. (1983b). Mitochrondrial codes and evolution. *Nature*, **301**, 19–20.

Jukes, T. H. (1993). The genetic code function and evolution. *Cellular and Molecular Biology Research*, **39**, 685–8.

Jukes, T. H. (1996). Oparin Medal Challenge. *ISSOL News Letter*, **23**, 20.

Jukes, T. H. (1997). Oparin and Lysenko. *Journal of Molecular Evolution*, **45**, 339–41.

Jukes, T. H., & Bhushan, V. (1986). Silent nucleotide substitutions and G + C content of some mitochrondrial and bacterial genes. *Journal of Molecular Evolution*, **24**, 39–44.

Jukes, T. H., Osawa, S., & Lehman, N. (1987). Evolution of anticodons: Variations in the genetic code. *Cold Spring Harbor Symposia on Quantitative Biology*, **52**, 769–76.

Jull, A. J. T., Courtney, C., Jeffrey, D. A., & Beck, J. W. (1998). Isotopic evidence for a terrestrial source of organic compounds found in Martian meteorites Allan Hills 84001 and Elephant Moraine 79001. *Science*, **279**, 366–9.

Kasting, J. F. (1993). Earth's early atmosphere. *Science*. **259**, 920–6.

Kauffman, S. A. (1993). *The origin of order: Self-organization and selection in evolution*. Oxford: Oxford University Press.

Keene, James D. (2001). Ribonucleoprotein infrastructure regulating the flow of genetic information between the genome and the proteome. *Proceedings of the National Academy of Sciences*, **98**, 7018–24.

Kemeny, J., Snell, J. L., & Knapp, A. W. (1976). *Denumerable Markov Chains*. New York: Springer Verlag.

Kerr, Richard. (2003a). Minerals cooked up in the laboratory call ancient microfossils into question. *Science*, **302**, 1134.

Kerr, Richard. (2003b). Eons of a cold, dry, dusty Mars. *Science*, **313**, 1037–8.

Kerr, Richard A. (1999). Early life thrived despite earthly travails. *Science*, **284**, 2111–13.

Kerr, Richard A. (2001). Putting a lid on life on Europa. *Science*, **294**, 1258–9.

Kerr, Richard A. (2002). Reversals reveal pitfalls in spotting ancient and E. T. life. *Science*, **296**, 1384–5.

Kerr, Richard A. (2004). No din of alien chatter in our neighborhood. *Science*, **303**, 1133.

Khinchin, A. I. (1953). The entropy concept in probability theory. *Uspekhi Matematicheskikh Nauk*, **VIII**, 3–20.

Khinchin, A. I. (1957). *Mathematical Foundations of Information Theory*. Trans. by R. A. Silverman & M. D. Friedman. New York: Dover Publications.

Kimberlin, R. H. (1982). Scrapie agent: Prions or virinos? *Nature*, **297**, 107–8.

Kinsella, Rhoda J., Fitzpatrick, David A., Creevey, Christopher J., & McInerney, James O. (2003). Fatty acid biosynthesis in *Mycobacterium tuberculosis:* Lateral gene transfer, adaptive evolution, and gene duplication. *Proceedings of the National Academy of Sciences*, **100**, 10320–5.

Kipling, Rudyard. (1902). *Just So Stories*. New York: Garden City Books.

Kiseleva, E. V. (1989). Secretory protein synthesis in Chironomus salivary gland cells is not coupledwith protein translocation across endoplasmic reticulum membrane. *FEBS Letters*, **257**, 251–3.

Kleine, T., Münker, C., Mezger, K., & Palme, H. (2002). Rapid accretion and early core formation on asteroids and the terrestrial planets from Hf-W chronometry. *Nature*, **418**, 952–5.

Klemke, M., Kehlenbach, R. H., & Huttner, Wieland B. (2001). Two overlapping reading frames in a single exon encode interacting proteins–a novel way of gene usage. *EMBO J.*, **20**, 3849–60.

Klug, W. S., & Cummings, M. R. (1986). *Concepts of Genetics.* second edition. Columbus, OH: Merrill.

Kolata, G. B. (1977). Overlapping genes: More than anomalies? *Science*, **196**, 1187–8.

Kolmogorov, A. N. (1933). Grundbegriffe der Wahrscheinlichkeitsrechnung. *Ergebnisse der Mathematik*, **2**, No. 3 Berlin: Springer-Verlag.

Kolmogorov, A. N. (1958) A new metric of invariants of transitive dynamical systems and automorphisms in Lebesgue spaces. *Dokl. Akad. Nauk SSSR*. **119**, 861–864.

Kolmogorov, A. N. (1965). Three approaches to the concept of the amount of information, *IEEE Problems on Information Transmission*, **1**, No. 1–7.

Koonin, Eugene V., Wolf, Yuri I., & Karev, Georgy P. (2002). The structure of the protein universe and genome evolution. *Nature*, **420**, 218–23.

Kortemme, Tanja, & Baker, David. (2002). A simple physical model for bindng energy hot spots in protein-protein complexes. *Proceedings of the National Academy of Sciences*, **99**, 14116–21.

Kozak, Marilyn. (2001). Extensively overlapping reading frames in a second mammilian gene. *EMBO Reports*, **2**, 768–9.

Krauss, Lawrence M., & Chaboyer, Brian. (2003). Age estimates of globular clusters in the milky way: Constraints on cosmology. *Science*, **299**, 65–9.

Kryukov, Gregory V., Castellano, Sergei, Novoselov, Sergey V., Lobanov, Alexey V., Zehtab, Omid, Guigó Roderic, & Gladyshev, Vadim N. (2003). Characterization of mamilian selenoproteomes. *Science*, **300**, 1438–43.

Kuchino, Y., Beier, H., Akita, N., & Nishimura S. (1987). Natural UGA suppresser glutamine tRNA is elevated in mouse cells infected with Moloney murine leukemia virus. *Proceedings of the National Academy of Sciences*, **84**, 2668–72.

Kullback, S. (1968). *Information Theory and Statistics.* New York: Dover Publications.

Kunisawa, T., Horimoto, K., & Otsuka, J. (1987). Accumulation pattern of amino acid substitutions in protein evolution. *Journal of Molecular Evolution*, **24**, 357–65.

Kurland, C. G., Canback, B., & Berg, Otto G. (2003). Horizontal gene transfer: A critical view. *Proceedings of the National Academy of Sciences*, **100**, 9658–62.

Labouygues, J.-M. (1984). The logic of the genetic code: Synonyms and optimality against effects of mutations. *Origins of Life*, **14**, 405–13.

Labouygues, J.-M., & Figureau, A. (1982), L'Origine et l'évolution du code génétique. *Reviews of Canadian Biology Esperiments*, **41**, 209–16.

Lacey, J. C., & Mullin, D. W. Jr. (1983). Experimental studies related to the origin of the genetic code and the process of protein synthesis: A review. *Origins of Life*, **13**, 3–42.

Lacourciere, Gerard M., & Stadman, Thressa C. (1999). Catalytic properties of selenophosphate synthetases: Comparison of selenocysteine-containing enzyme from *Haemophilus influenzae* with the corresponding cysteine-containing enzyme from *Escherichia coli*, *Proceedings of the National Academy of Sciences*, **96**, 44–8.

Lahav, Noam. (1999). *Biogenesis, Theories of Life's Origin.* New York: Oxford University Press.

Lake, James A., Jain, Ravi, & Rivera, Maria C. (1999). Mix and match in the tree of life. *Science*, **283**, 2027–8.

Lancaster, Peter, & Tismenetsky, Miron. (1985). *The Theory of Matrices*. second edition. San Diego: Academic Press.

Landauer, R. (1986). Computation: A Fundamental Physical View, in *Maxwell's Demon, Entropy, Information, Computing*. Harvey S. Leff Andrew F. Rex, eds. Princeton Series in Physics. Princeton, NJ: Princeton University Press.

Landauer, R. (2000). Irreversibility and heat generation in the computing process. *IBM Journal of Research and Development*, **44**, 261–9.

Lang, David A., Smith, G. David, Courseille, Chritian, Précigoux, Gilles, & Hospital, Michel. (1991). Monoclinic uncomplexed double-stranded, antiparallel, left-handed $\beta^{5.6}$-helix ($\uparrow\downarrow\beta^{5.6}$) structure of gramicidin A: Alternative patterns of helical association and deformatin. *Proceedings of the National Academy of Sciences*, **98**, 5345–9.

Langkjaer, Rikke B., Cliften, Paul F., Johnston, Mark, & Piskur, Jure. (2003). Yeast genome duplication was followed by asynchronous differentiation of duplicated genes. *Nature*, **421**, 848–52.

Lawrence, Jeffrey G., & Ochman, Howard. (1998). Molecular archaeology of the *Escherichia coli* genome. *Proceedings of the National Academy of Sciences*, **95**, 9413–17.

Lazcano, A. (1997). Chemical evolution and the primitive soup: Did Oparin get it all right? *Journal of Theoretical Biology*, **184**, 219–23.

Lazcano, Antonio, & Miller, Stanley L. (1994). How long did it take for life to begin and evolve to cyanobacteria? *Journal of Molecular Evolution*, **39**, 546–54.

Lee, Siu Sylvia, Kennedy, Scott, Tolonen, Andrew. C., & Ruvkun, Gary. (2003). DAF-16 target genes that control *C. elegans* life-span and metabolism. *Science*, **300**, 644–7

Lehman, N., & Jukes, T. H. (1988). Genetic code development by stop codon takeover. *Journal of Theoretical Biology*, **15**, 203–14.

Lehn, Jean-Marie. (2002a). Toward self-organization and complex matter. *Science*, **295**, 2400–3.

Lehn, Jean-Marie. (2000b). Toward complex matter: Supramolecular chemistry and self-organization. *Proceedings of the National Academy of Sciences*, **99**, 4763–8.

Leinfelder et al. (1988). Gene for a novel tRNA species that accepts L-serine and cotranslationally inserts selenocysteine. *Nature*, **331**, 723–5.

Leinfelder, W., Zehelein, E., Mandradn-Bertholot, M.-A., & Bock, A.. (1988). Gene for a novel tRNA species that accepts L-serine and cotranslationally inserts selenocystein. *Nature*, **331**, 723–5.

Lenski, R. E., Ofria, C. Collier, T., & Adami, C. (1999). Genome complexity, robustness, and genetic interactions in digital organisms. *Nature*, **400**, 661–4.

Levitt, M., & Gerstein, M. (1998). A unified statistical framework for sequence comparison. *Proceedings of the National Academy of Sciences*, **95**, 5913–20.

Levy, M., & Miller, S. L. (1998). The stability of the RNA bases: Implications for the origin of life. *Proceedings of the National Academy of Sciences*, **95**, 7933–8.

Li, M., & Tzagoloff, A. (1979). Assembly of the miotochondrial membrane system: Sequence of yeast mitochondrial valine and an unusual threonine tRNA gene. *Cell*, **18**, 311–13.

Li, Ming, & Vitányi, Paul. (1997). *An Introduction to Kolmogorov Complexity and Its Applications*. second edition. New York: Springer Verlag.

Liang, N., Peilak, G. J., Johnson, J. A. Smith, M., & Hoffman, B. M. (1987). Yeast cytochrome c with phenylanine or tyrosine at position 87 transfers electrons to (zinc cytochrome c peroxidase) + at a rate ten thousand that of serine -87 or glycine-87 variants. *Proceedings of the National Academy of Sciences*, **84**, 1249–52.

Liebman, Susan. (2002). Progress toward an ultimate proof of the prion hypothesis, *Proceedings of the National Academy of Sciences*, **99**, 9098–100.

Lindley, D. V. (1965). *Introduction to Probability and Statistics, Part I*. Cambridge, UK: Cambridge University Press.

Lineweaver, Charles H. (1999). A younger age for the universe. *Science*. **284**, 1503–7.

Löb, W. (1913). Über das Verhalten des Formids unter der Wirkung der stillen Entladung: Ein Beitrag zur Frage der Stickstoff-Assimilation. *Berichte der deutsche chemische Gesellschaft*, **46**, 684–97.

Löb, Walther. (1904). Zur Kenntnis der Assimilation der Kohlensäure. *Ber.*, **37**, 3539–96.

Löb, Walther. (1905). Zur Kenntnis der Assimilation der Kohlensäure. *Zeitschrift für Elektrochemie*, **11**, 745–63.

Löb, Walther. (1906). Studien über die chemische Wirkung der stillen elektrischen Entladung. *Zeitschrift für Elektrochemie*, **12**, 282–313.

Löb, Walther. (1907). Zur chemischen Theorie der alkoholischen Gärung. *Zeitschrift für Elektrochemie*, **13**, 511–18.

Löb, Walther. (1908a). Die Einwirkung der stillen elektrischen Entladung auf feuchtes Methan. *Ber.*, **41**, 87–90.

Löb, Walther. (1908b). Über die Einwirkung der stillen Entladung auf feuchten Stickstoff und feuchtes stickoxyd. *Zeitschrift für Elektrochemie*, **14**, 556–64.

Löb, Walther. (1908c). Über die Bildung von Wasserstoffperoxyd durch stille elektrische Entladung. *Ber.*, **41**, 1517–18.

Löb, Walther. (1909a). Über die Bildung von Buttersäure aus Alkohol unter dem Einfluß der stillen Entladung. *Biochemische Zeitschrift*, **20**, 126–35.

Löb, Walther. (1909b). Über die Aufnahme des Stickstoffs durch Alkohol unter dem Einfluß der stillen Entladung. *Biochemische Zeitschrift*, **20**, 136–42.

Löb, Walther. (1912). Über das Verhalten der Stärke unter dem Einflüß der stillen Entladung. *Biochemische Zeitschrift*, **46**, 121–4.

Löb, Walther. (1914). Über die Einwirkung der stillen Entladung auf Stärke und Glykokoll. *Biochemische Zeitschrift*, **60**, 285–96.

Löb, Walther. (1915). Das Verhalten des Rohrzuckers bei der stillen Entladung. *Biochemische Zeitschrift*, **69**, 36–8.

Löb, Walther, & Sato, A. (1915). Zur Frage der Elektrokultur. *Biochemische Zeitschrift*, **69**, 1–34.

Loeb, Jacques. (1906). *The Dynamics of Living Matter*. London: Macmillan.

Loeb, Jacques. (1912). The mechanistic conception of life. In *The Mechanistic Conception of Life*. Chicago: University of Chicago Press.

Loeb, Jacques. (1916). *The Organism as a Whole*. New York: G. P. Putnam's Sons.

Loeb, Jacques. (1924). *Proteins and the Theory of Colloidal Behavior*. London: McGraw-Hill.

Lovejoy, Arthur O. (1936). *The Great Chain of Being*. Cambridge. MA: Harvard University Press.

Lynch, Michael, & Conery, Johns. (2000). The evolutionary fate and consequences of duplicate genes. *Science*, **290**, 1151–5.

Lynch, Michael. (2002). Gene duplication and evolution. *Science*, **297**, 945–7.

Macino, G. Coruzzi, G. F. G., Li, M., & Tagoloff, A. (1979). Use of UGA terminator as a tryptophan codon in yeast. *Proceedings of the National Academy of Sciences*, **76**, 3784–5.

Maddelein, Marie-Lise, Dos Rios, Suzana, Duvezin-Caubet, Coulary-Salin & Saupe, Bénédicte, & Saupe, Sven J. (2002). Amyloid aggregates of the HET-s proten are infectious. *Proceedings of the National Academy of Sciences*, 7402–7.

Maddox, Brenda. (2002). *Rosalind Franklin: The dark lady of DNA*. New York: Harper Collins.

Maeshiro, Tetsuya, & Kimura, Masayuki. (1998). The role of robustness and changebililty on the origin and evoluton of genetic codes. *Proceedings of the National Academy of Sciences*, **95**, 5088–93.

Makous, Walter. (2000). Limits to our knowledge. *Science*, **287**, 1399.

Malin, Michael C., & Edgett, Kenneth S. (2003). Evidence for persistent flow and aqueous sedimentation on early Mars. *Science*, **302**, 1931–4.

Mann, S., Sparks, N. H. C., Frankel, R. B., Bazylindkie, D. A., & Holger, W. (1990). Biomineralization of ferrimagnetic greliglite (Fe_3S_4) and ironpyrite (FeS_2) in a magnetostsatic bacterium. *Nature*, **343**, 258–61.

Marcus, Marvin, & Minc, Hendryk. (1984). *A Survey of Matrix Theory and Matrix inequalities*. New York: Dover.

Margoliash, E., Fitch, W. M., Markowitz, E., & Dickerson, R. E. (1972). In *Structure and Function of Oxidation-Reduction Enzymes*. Wenner-Gren Symposium 1970 (A. Akeson and A. Ehrenberg, Eds.), Pergamon Press, Oxford, pp. 5–17.

Margulis, L. (1970). *Origin of eucaryote cells*. New Haven, CN: Yale University Press.

Marks, C. B., Naderi, H., Kosen, P. A. Kuntz, I. D., & Anderson, A. (1987). Mutants of bovine pancreatic trypsin inhibitor lacking cysteines 14 and 38 can fold properly. *Science*, **235**, 1370–3.

Martin-Löf, P. (1966). The definition of randomness. *Information and Control*, **9**, 602–19.

Martinac, Boris, & Hamill, Owen P. (2002). Gramicidin A channels switch between stretch activation and stretch inactivation depending on bilayer thickness. *Proceedings of the National Academy of Sciences*, **99**, 4308–12.

Maynard Smith, J. (2000). The Concept of Information in Biology. *Philosophy of Science*, **67**, 177–94.

Maynard Smith, John. (1999) .Too good to be true. *Nature*, **400**, 223.

Mayr, E. (1982). *The Growth of Biological Thought*. Cambridge MA: The Belknap Press of Harvard University Press.

Mayr, E. (1988). Introduction, and Is biology an autonomous science? In *Toward a New Philosophy of Biology* (pp. 1–7 and 8–23). Cambridge, MA: Harvard University Press.

McKay, Christopher P. (1991). Urey Prize Lecture: Planetary evolution and the origin of life. *Icarus*, **91**, 93–100.

McMahon, Robert J. (2003). Chemical reactions involving quantum tunneling. *Science*, **299**, 867–70.

McMillan, B. (1953). The basic theorems of information theory. *Annals of Mathematical Statistics*, **24**, 196–219.

Medvedev, Zhores A. (1969). *The Rise and Fall of T. D. Lysenko*. Trans. I. Michael Lerner. New York: Columbia University Press.

Melosh, J. (2001). A new model Moon. *Nature*, **412**, 694–5.

Mendel, Gregor J. (1866). *Versuche über pflantzen-hybriden, Verhandlungen Naturforschen Brünn.*

Merbach, Marlis A., Merbach, Dennis A., Maschwitz, Ulrich, Booth, Webber E., Fiala, Brigitte, & Zizka, Georg. (2002). Carnivorous plants: Mass march of termites into the deathly trap. *Nature*, **415**, 36–7.

Meyer, Axel. (2003). Molecular evolution: Duplication, duplication. *Nature*, **421**, 31–2.

Miller, S. L., & Orgel, L. E. (1974). *The Origins of Life on Earth.* Englewood Cliffs, NJ; Prentice-Hall.

Miller, S. L., Schopf, J. W., and Lazcano, A. (1997). Oparin's "Origin of Life": Sixty years later. *Journal of Molecular Biology*, **44**, 351–3.

Miller, Stanley L. (1953). A production of amino acids under possible primitive Earth conditions. *Science*, **117**, 528–9.

Miller, Stanley L. (1955). Production of some organic compounds under possible primitive earth conditions. *Journal of the American Chemical Society*, **77**, 2351–61.

Mitchison, G. F., & Durbin, R. (1995). Tree-based maximal likelihood substitution matrices and hidden Markov models. *Journal of Molecular Evolution*, **41**, 1139–51.

Miyakawa, Shin, Yamanashi, Hiroto, Kobayashi, Kensei, Cleaves. H. James, & Miller, Stanley L. (2002). Prebiotic synthesis from CO atmospheres: Implications for the origins of life. *Proceedings of the National Academy of Sciences*, **99**, 14628–31.

Mizutani, T., & Hitaka, T. (1988). The conversion of phosophoserin residues to selenocysteine residues on an opal suppresser tRNA and casein. *Federation of European Biochemical Societies Letters*, **232**, 243–8.

Mojzsis, S. J., Kishnamurthy, R., & Arrhenius, G. (1999). Before RNA and After: Geological and Geochemical Constraints on Molecular Evolution. In *The RNA World: The Nature of Modern RNA* suggests a prebiotic RNA (second edition), Raymond F. Gesteland, ed. (pp. 1–47). Cold Spring Harbor Laboratory Press. Cold Spring Harbor, New York.

Mojzsis, Stephen J. Harrison, T. Mark (2002). Origin and significance of Archean Quartzose Rocks at Akilia, Greenland. *Science*, **298**, 917a.

Mojzsis, Stephen J., Harrison, T. Mark & Pidgeon. (2001). Oxygen-isotope evidence from ancient zircons for liquid water at the Earth's surface 4,300 Myr ago. *Nature*, **409**, 178–81.

Montoya, J., Ojla, D. & Attardi, G. (1981). Distinctive features of the 5′ terminal sequences of the human mitochondrial mRNAs. *Nature*, **290**, 465–70.

Moore, Richard, & Purugganan, Michael D. (2003). The early stages of duplicate gene evolution. *Proceedings of the National Academy of the Sciences*, **100**, 15682–7.

Moran, P. A. P. (1986). *An Introduction to Probability Theory.* Clarendon Press, Oxford University Press, Oxford.

Morowitz, Harold J., Kostelnik, Jennifer, D., Yang, Jeremy, & Cody, George D. (2000). The origin of intermediary metabolism. *Proceedings of the National Academy of Sciences*, **97**, 7704–8.

Morrison, David. (2001). The NASA astrobiology program. *Astrobiology*, **1**, 3–13.

Morton, Oliver. (2003). Mars revisited. *National Geographic*, January 2004.

Mullenbach, G., Tabrizi, A., Irvine, B. D., Bell, G. I., & Hallewell, R. A. (1987). Sequence of a cDNA coding for human glutathione peroxidase confirms TGA encodes active site selenocysteine. *Nucleic Acids Research*, **15**, 5484.

Muller, H. J. (1966). The gene material as the initiator and the organizing basis of life. *American Naturalist*, **100**, 493–517.

Münker, Carsten, Pfänder, Weyer, Stefan, Büchl, Kleine, Thorsten, & Mezger, Klaus. (2003). Evolution of planetary cores and the Earth–Moon system from Nb/Ta systematics. *Science*, **301**, 84–7.

Murphy, Coleen T., McCarroll, Steven A., Bargman Cornelia I., Fraser, Amdrew. Kamath, Revi S., Ahringer, Julie, Hao Li, & Kenyon, Cynthia. (2003). Genes that act downstream of DAF-16 to influence the lifespan of *Caenorhatiditis elegans*. *Nature*, **424**, 277–84.

Mushegian, A. R., & Koonin, E. V. (1996). A minimal gene set for cellular life derived by comparison of complete bacterial genomes. *Proceedings of the National Academy of Sciences*, **93**, 10268–73.

Muto, A., Yamao, F., Hori, H., & Osawa, S. (1985). Gene organization of *Mycoplasma caprolium*. *Advances in Biophysics*, **21**, 49–56.

Nelson, Kevin E., Levy, Matthew, & Miller, Stanley, L. (2000). Peptide nucleic acids rather than RNA may have been the first genetic molecule. *Proceedings the National Academy of Sciences*, **97**, 3868–71.

Nisbet, E. G., & Sleep, N. H. (2001). The habitat and nature of early life. *Nature*, **409**, 1083–91.

Noren, C. J., Anthony-Cahill, S. J., Griffith, M. C., & Schult, P. G. (1989). A general method for site-specific incorporation of unnatural amino acids into proteins. *Science*, **244**, 182–8.

Oba, Takanori, Andachi, Yoshiki, Muto, Akira, & Osawa, Syozo. (1991). CGG: An unassigned or nonsense condon in *Mycoplasma capricolum*. *Proceedings of the National Academy of Sciences*, **88**, 921–5.

Ohno, Susumu. (1970). *Evolution by gene duplication*. Berlin: Springer Verlag.

Oparin, A. I. (1924). *Proiskhozdenic Zhizny*. Moscow: Izd Moskovski Rabochii. [Published in English. trans. in J. D. Bernal. (1967). *The origin of life*. London: Weidenfeld & Nicolson.]

Oparin, A. I. (1938). *Origin of life*. New York: Macmillan.

Oparin, A. I. (1957). *The Origin of Life on Earth*. third revised edition, trans. Ann Synge. Edinburgh: Oliver & Boyd.

Oparin, A. I. (1964). *LIFE: Its Nature, Origin and Development*. trans. from the Russian by Ann Synge. New York: Academic Press.

Orgel, L. E. (1968). Evolution of the genetic apparatus. *Journal of Molecular Biology*, **38**, 381–93.

Orgel, L. E. (1986). RNA catalysis and the origin of life. *Journal of Theoretical Biology*, **123**, 127–49.

Orgel, Leslie E. (2000). Self-organizing biochemical cycles. *PNAS*, **97**, 12503–7.

Ornstein, D. S. (1974). Ergodic Theory, Randomness, and Dynamical Systems. *Yale Mathematical Monographs*, **5**. New Haven, CT: Yale University Press.

Ornstein, Donald. (1970). Bernoulli shifts with the same entropy are isomorphic. *Advances in Mathematics*, **4**, 337–52.

Ornstein, Donald S. (1989). Ergodic theory, randomness and "chaos." *Science*, **243**, 182–7.

Osawa, S. (1995). *Evolution of the Genetic Code*. Oxford: Oxford University Press.

Osawa, S. et al. (1990). Evolutionary changes in the genetic code. *Proc. R. Lond. B.*, **241**, 19–28.

Osawa, S., Jukes, T. H., Muto, A., Yamao, F., Ohama T., & Andachi, Y. (1987). Role of directional mutation pressure in the evolution of the eubacterial genetic code. *Cold Spring Harbor Symposia on Quantitative Biology*, **52**, 777–89.

Osawa, S., Jukes, T. H., Watanabe, K., & Muto, A. (1992). Recent Evidence for Evolution of the Genetic Code. *Microbiological Reviews*, **56**, 229–64.

Overman, Dean L., & Yockey, Hubert P. (2001). Information, Algorithms and the Unknowable Nature of Life's Origin. *The Princeton Theological Review*, **VIII**, No. 4.

Overman, Dean L. (1997). *A Case against Accident and Self-Organization*. Lanham, MD: Rowman & Littlefield Publishers.

Pais, J., (1982). *"Subtle is the Lord": The Science and the Life of Albert Einstein*. Oxford: Clarenden Press.

Paley, William. (1802). Natural Theology; or, Evidences of Existence and attributes of the Deity, An University of Michigan Humanities Text Initiative Ann Arbor MI.

Parthia, R. K. (1962). A statistical study of randomness among the first 10,000 digits of π. *Mathematics of Computation*, **16**, 188–97.

Pasteur, Louis. (1922). Dissymétry moléculaire. In *Oeuvres de Pasteur, Tome premier*. Paris: Masson.

Pasteur, Louis. (1848). Sur les relations qui peuvent exister entre la forme cristalline, la composition chimique et le sens de la polarization rotatoire. *Annales de Chimie Physique*, **24**, 442–59.

Paul, Ligi, Ferguson, Donald J. Jr., & Krzycki, Joseph A. (2000). The trimethylamine methyltransferase gene and multiple dimethylamine methyltransferase genes of *Methanoscarcina barkeri* contain in-frame and read-through amber codons. *Journal of Bacteriology*, **182**, 2520–9.

Pauling, Linus, & Corey, R. B. (1953). *Proceedings of the National Academy of Sciences*, **39**, 84.

Pauling, Linus. (1987). Schrödinger's contribution to chemistry and biology. In *Schrödinger: Centenary of a Polymath*. Cambridge, UK: Cambridge University Press.

Pauling, Linus, & Corey, R. B. (1953). *Nature*, **171**, 346.

Pauly, P. J. (1987). *Controlling Life: Jacques Loeb and the Engineering Ideal in Biology*. Oxford: Oxford University Press.

Pavesi, Angelo, De Iaco, Betina, Granero, Maria Ilde, & Porati, Alfredo. (1997). On the informational content of overlapping genes in prokaryotic and eukaryotic viruses. *Journal of Molecular Evolution*, **44**, 625–31.

Pelc, S. R., & Welton, M. G. E. (1966). Steriochemical relationship between coding triples and amino-acids. *Nature*, **209**, 868–70.

Pennisi, Elisabeth. (2003). DNA's cast of thousands. *Science*, **300**, 282–5.

Penny, David. Evolutionary biology: Our relative genetics. *Nature*, **427**, 208–11. (2004)

Penrose, R. (1989). *The Emperor's New Mind*. Oxford: Oxford University Press.

Peretz, David, Williamson, R. Anthony, Kaneko, Kiotoshi, Vergara, Julie, Leclerc, Estellle, Schmit-Ulms, Gerold, Melhorn, Ingrid R., Legname, Guiseppe, Wormald, Mark R., Rudd, Pauline M., Dwek, Raymond, A., Burton, Dennis R., & Prusiner, Stanley B. (2001). Antibodies inhibit prion propagation and clear cell cultures of prion infectivity. *Nature*, **412**, 739–43.

Perlwitz, M. D., Burks, C., & Waterman, M. S. (1988). Pattern Analysis of the Genetic Code. *Advances in Applied Mathematics*, **64**, 9, 7–21.

Perron, O. (1907). Zur theorie der matrizen. *Mathematischen Annalen*, **64**, 248–63.

Petersen, Karl. (1983). *Ergodic Theory*. Cambridge, UK: Cambridge University Press.

Petz, Dénes. (2001). Entropy, von Neumann and the von Neumann entropy. In *John von Neumann and the Foundations of Quantum Physics*, M. Rédeland & M. Stöltzner, eds. Norwell, MA: Kluwer Academic.

Pincus, S., & Singer, B. H. (1996). Randomness and degrees of irregularity. *Proceedings the National Academy of Sciences*, **93**, 2083–88.

Pincus, Steve, & Kalman, Rudolf. (1997). Not all (possible) "random" sequences are created equal. *Proceedings of the National Academy of Sciences*, **94**, 3513–18.

Pincus, Steve, & Singer, Burton H. (1998). A recipe for randomness. *Proceedings of the National Academy of Sciences*, **95**, 10367–72.

Ponce de León, Marcia S., & Zollikofer, Christopher P. (2001). Neanderthal cranial ontogeny and its implications for late hominid diversity. *Nature*, **412**, 534–8.

Ponnamperuma, C. (1983). Cosmochemistry and the origin of life. In *Cosmochemistry and the Origin of Life*. C. Ponnamperuma, ed. Dordrecht: Reidel.

Popa, Radu. (1997). A sequential scenario for the origin of biological chirality. *Journal of Molecular Evolution*, **44**, 121–7.

Popper, K. R. (1990). Pyrite and the origin of life. *Nature*, **344**, 387.

Prusiner, S. B. (1982). Novel proteinaceous infectious particles cause scrapie. *Science*, **216**, 136–44.

Prusiner, Stanley B. (1998). Prions. *Proceedings of the National Academy of Sciences*, **95**, 13363–83.

Rather, Lelland J. (1958). *Disease, Life and Man: Selected Essays by Rudolf Virchow*. Trans. with an introduction by Lelland J. Rather. Stanford: Stanford University Press.

Raup, D. M., & Valntine, J. W. (1983). Multiple origins of life. *Proceedings of the National Academy of Sciences*, **80**, 2981–4.

Reddy, V. B. Thermmappaya, B., Dhar, R., Subramamian, K. N., Zain, B. S., Pan, J., Ghosh, P. K. Celma, M. L., & Weissman, S. M. (1978). The genome of simian virfus 40. *Science*, **200**, 494–502.

Reidhaar-Olson, J. F., & Sauer, R. T. (1988). Combinatorial cassette mutagenesis as a probe of the informational content of protein sequences. *Science*, **241**, 55–7.

Rasmussen, Steen, Chen, Liaohi, Deamer, David, Krakauer, David C. Packard, Norman H., Stadler, Peter F., & Bedau, Mark A. (2004). Transitions from Nonliving to Living Matter. *Science*, **303**, 963–5.

Rich, A. (1962). On the problems of evolution and bioichemical transfer. In *Horizons in Biochemistry*, M. Kasha & B. Pullman, eds. (pp. 103–26). New York: Academic Press.

Richardson, J. S., & Richardson, D. C. (1988). Amino acid preferences for specific locations α helices. *Science*, **240**, 1648–52.

Romanek, Christopher S., Clement, Simon J., Chillier. Xavier D. F., Maechling, Claude, R., & Zare, Richard N. (1996). Search for past life on Mars: Possible relic biogenic activity in Martian meteorite ALH84001. *Science*, **273**, 924–30.

Sagan, C. (1973). Ultraviolet selection pressure on the earliest organisms. *Journal of Theoretical Biology*, **39**, 195–200.

Sagan, C., & Mullin, G. (1972). Earth and Mars: Evolution of atmospheres and surface temperatures. *Science*, **177**, 52–6.

Sagan, Carl. (1985). *Contact: A novel*. New York: Simon & Schuster

Salzberg, Stephen J., White, Owen, Peterson, Jeremy, & Eisen, Jonathan A. (2001). Microbial genes in the human genome: Lateral transfer or Gene loss? *Science*, **292**, 1903–6.

Sanger, F., Air, G. M., Barrell, B. G., Brown, N. L. Coulson, A. R. Fiddes, J. C., Hutchinson, C. A., III, Solcombe, P. M., & Smith, M. (1977). Nucleotide sequences of bacteriophage ΦX174. *Nature*, **268**, 687–95.

Sassetti, Christopher, M., Boyd, Dana H., & Rubin, Eric J. (2001). Comprehensive identification of conditionally essential genes in mycobacteria. *Proceedings of the National Academy of Sciences*, **98**, 12712–17.

Schidlowski, M. (1983). Biologically mediated isotope fractionations: Biochemistry, geochemical significance and preservation in the Earth's oldest sediments. In *The Early History of the Earth*, B. F. Windley, ed. London: John Wiley & Sons.

Schidlowski, M. (1988). A 3,800-million-year isotope record of life from carbon in sedimentary rocks. *Nature*, **333**, 313–18.

Schidlowski, Manfred. (2002). Sedimentary carbon isotope archives as recorders of early life: implications for extraterrestrial scenarios. In *Fundamentals of life*, Pályi Gyala, Claudia Zucchi, & Luciano Caglioti, eds. Paris: Elsevier.

Schlesinger, Gordon, & Miller, Stanley L. (1983). Prebiotic synthesis in atmospheres containing CH_4, CO and CO_2. I. Amino Acids. *Journal of Molecular Evolution*, **19**, 376–82.

Schneider, T. D. Stormo, G. D. Gold, L., & Ehrenfeucht, A. (1986). Information content of binding sites on nucleotide sequences. *Journal of Molecular Biology*, **188**, 415–31.

Schneider, Thomas D. (2000). Evolution of biological information. *Nucleic Acids Research*, **28**, 2794–9.

Schoenberg, Ronny, Kamber, Balz S., Collerson, Kenneth D., & Moorbath, Stephen. (2002). Tungsten isotope evidence from -3.8-Gyr metamorphosed sediments for early meteorite bombardment of the Earth. *Nature*, **418**, 403–5.

Schopf, William J. (1999). *Cradle of Life*. Princeton, NJ: Princeton University Press.

Schopf, William J., Kudryavtsev, Anatolly B., Agresti, David, G., Wdowiak, Thomas J., & Czaja, Andrew D. (2002). Laser-Raman imagery of Earth's earliest fossils. *Nature*, **416**, 73–81.

Schröder, E. (1870). *Z. Math. Phys.*, **15**, 361–76.

Schrödinger, Erwin. (1992). *What is Life? With Mind and Matter and Autobiographical Sketches*. Cambridge, UK: Cambridge University Press.

Schrödinger, Erwin. (1987). *Centenary of a Polymath*. Ed. C. W. Kilmister. Cambridge, UK: Cambridge University Press.

Schulz, A. S., Shmoys, D. B., and Williamson, D. P. (1994). Approximation algorithms, *Proceedings of the National Academy of Sciences*, **94**, 12734–5.

Schuster, Peter. (2000). Taming combinatorial explosion, *Proceedings of the National Academy of Sciences*, **97**, 7678–80.

Schwemmier W. *Reconstruction of Cell Evolution*, Boca Raton, FL: CRC Press, p. 248.

Segrè, Gino. (2000). The Big Bang and the genetic code. *Nature*, **404**, 437–8.

Segura, Teresa L., Toon, Owen B., Colaprete, Anthony, & Zahnle, Kevin. (2002). Environmental effects of large impacts on Mars. *Science*, **298**, 1997–80.

Selkoe, Dennis J. (2003). Folding proteins in fatal ways. *Nature*, **426**, 900–4.

Seneta, E. (1973). *Non-Negative Matrices. An Introduction to Theory and Applications*. London: George Allen & Unwin.

Shannon, C. E. (1948). A mathematical theory of communication. *Bell System Technical Journal*, **27**, 379–424, 623–56.

Shannon, C. E. (1951). Prediction and entropy of printed English. *Bell System Technical Journal*, **30**, 50–64.

Shapiro, R. (1999). Prebiotic cytosine synthesis: A critical analysis and implications for the origin of life. *Proceedings of the National Academy of Sciences*, **96**, 4396–401.

Shaw, J. M., Walker, J. E., Northrop, F. D., Barrell, B. G., Godson, G. N., & Fiddes, J. C. (1978). Gene K, a new overlapping gene in bacteriphage G4. *Nature*, **272**, 510–15.

Shields, P. C. (1973). *The Theory of Bernoulli Shifts*. Chicago: University of Chicago Press.

Shiguru, Ida, Canup, Robin M., & Stewart, Glen, R. (1997). Lunar accretion from an impact generated disk. *Nature*, **389**, 353–7

Shklovskii, I. S., & Sagan, Carl. (1966). *Intelligent Life in The universe*. New York: Dell.

Shklovsky, Iosif. (1991). *Five Billion Vodka Bottles to the Moon*. Trans. and adapted by Mary Fleming Zirin & Harold Zirin, New York: W. W. Norton.

Shmulevitz, Maya, Yameen, Zareen, Dawe, Sandra, Shou, Jingyun, O'Hara, David, Holmes, Ian, & Duncan, Roy. (2002). *Journal of Virology*, **76**, 609–618.

Sibler, A.-P., Dirheimer, G. & Cover, T. M. (1981). Nucleotide sequence of a yeast mitochondrial threonine-tRNA able to decode the C-U-N leucine codons. *Federation of European Biochemical Society Letters*, **132**, 344–8.

Sillén, I. G. (1965). Oxidation state of Earth's ocean and atmosphere. *Arkiv för Kemi*, **24**, 431–56.

Simpson, G. G. (1964). The nonprevalence of humanoids. *Science*, **143**, 769–75.

Sinai, Ya. G. (1959). The notion of entropy of a dynamical system. *Dokl. Akad. Nauk.*, **125**, 768–71.

Smallwood, Hugh, & Urey, H. C. (1928). An attempt to prepare triatimic hydrogen. *Journal of the American Chemical Society*, **50**, 620–6.

Smith, J. V. (1998). Biochemical evolution. I. Polymerization on integral, organophilic silica surfaces of dealuminated zeolites and feldspars. *Proceedings of the National Academy of Sciences*, **95**, 3370–5.

Smith, J. V. (1999). Biochemcal evolution III: Polymerization on organophilic silica-rich surfaces, crystal-chemical modeling, formation of first cells, and geological clues. *Proceedings of the National Academy of Sciences*, **96**, 3479–85.

Smith, John Maynard, & Szathmáry, Eörs. (1995). *The Major Transitions in Evolution*. Oxford: W. H. Freeman.

Smith, John Maynard, & Szathmáry, Eörs. (1999). *The Origins of Life*. Oxford: Oxford University Press.

Smith, L., Davies, H., Reichlin, M., & Margoliash, H. J. (1989). Separate oxidase and reductase reaction sites on cytochrome c demonstrated with purified site-specific antibodies. *Journal of Biological Chemistry*, **248**, 237–43.

Smith, M., Brown, G. M., Air, G. M., Barrell, B. G., Coulson, A. R., Hutchinson, C. A., III, & Sanger, F. (1977). DNA sequence at the C tremini of the overlapping genes A and B in bacteriophage ΦX174. *Nature* **265**, 702–5.

Smullyan, R. M. (1992). *Gödel's Incompleteness Theorems*. Oxford: Oxford University Press.

Solomonoff, T. M. (1964). A formal theory of inductive inference. *Information and Control*, **7**, 1–22, 224–54.

Sonneborn, T. M. (1965). Degeneracy of the Genetic Code: Extent, Nature and Genetic Implications. In *Evolving Genes and Proteins*, V. Bryson and Henry J. Vogel, eds. New York: Academic Press.

Sowerby, Stephen J., Cohn, Corey A., Heckl, Wolgang M. & Holm, Nila G. (2001) Diferential adsorption of nucleic acid bases: Relevance to the origin of life. *Proceedings of the National Academy of Sciences*. **98**, 820–822.

Sowerby, Stephen, & Heckl, Wolfgang M. (1998). The role of self-assembled monolayers of the purine and pyrimidine bases in the emergence of life. *Origin of Life and Evolution of the Biosphere*, **28**, 283–310.

Sowerby, Stephen J., Mörth, C.-M., & Holm, N. G. (2001). Effect of temperature on the adsorbtion of adenine. *Astrobiology*, **1**, 481–7.

Sowerby, Stephen J., & Peterson, George B. (2002). Life before RNA. *Astrobiology*, **2**, 231–8.

Sowerby, Stephen J., Peterson, G. B., & Holm, N. G. (2002). Primordial coding of amino acids by adsorbed purine bases. *Origin of Life and Evol. Biosphere*, **32**, 35–46.

Soyfer, Valery N. (1989). New Light on the Lysenko era. *Nature*, **339**, 415–20.

Soyfer, Valery N. (1994). *Lysenko and the Tragedy of Soviet science*. New Brunswick, NJ: Rutgers University Press.

Spencer, C. A., Gietz, R. D., & Hodgetts, R. B. Overlapping transcription units in the dopa decarboxylase region of *Drosophila*. *Nature*, **322**, 279–81. 1986

Srinivasan, Gayathri, James, Carey M., & Krzycki. (2002). Pyrrolysine encoded by UAG in Archaea: Charging of a UAG-decoding specialiazed tRNA. *Science*, **296**, 1459–62.

Stevens, Eric R., Esguerra, Manuel, Kim, Paul M., Newman, Eric A., Snyder, Solomon H., Zahs, Kathleen R., & Miller, Robert F. (2003). *Proceedings of the National Academy of Sciences*, **100**, 6789–94.

Stoneham, R. G. (1965). A study of 60,000 digits of the transcendental "e." *American Mathematical Monthly*, **72**, 483–500.

Strait, B., & Dewey, G. (1996). The Shannon information entropy of protein sequences. *Biophys. J.*, **71**, 148–55.

Stribling, R., & Miller S. L. (1987). Energy yields for hydrogen cyanide and formaldehyde syntheses: The HCN and amino acid concentration in the primitive ocean. *Origins of Life*, **17**, 2261–73.

Sunde, R. A., & Evenson, J. K. (1987). Serine incorporation into the selenocysteine moiety of glytathione peroxidase. *Journal of Biological Chemistry*, **262**, 933–7.

Suppes, P. (1972). *Axiomatic set theory*. New York: Dover Publications.

Surridge, Christopher. (2002). Plant science: On the slide. *Nature*, **420**, 753.

Swift, Jonathan. (1726). *Gulliver's travels*. Reprinted by Barnes & Noble Books, New York 1995.

Syvanen, M., & Kado, C. I. (1998). *Horizontal Gene Transfer*. London: Chapman & Hall.

Szathmáry, Eörs, & Maynard Smith, John. (1997). From replicators to reproducers: The first major transitions leading to life. *Journal of Theoretical Biology*, **187**, 555–71.

Szekely, M. (1977). *Phi X 174* sequenced. *Nature*, **265**, 685.

Szekely, M. (1978). Triple overlapping genes. *Nature*, **272**, 492.

Szostak, Jasck W., Bartel, David P., & Liusi, Luigi. (2001). Synthesizing life. *Nature*, **409**, 387–90.

Temin, H, M., & Mizutani, S. (1970). RNA-dependent DNA polmerase in virons of Rous sarcoma virus. *Nature*, **226**, 1211–13.

Thomas, Paul J., Chyba, Christopher F., & McKay, Christopher F., eds. (1997). *Comets and the Origin and Evolution of Life*. New York, Heideberg. Springer Verlag.

Thomas-Keprta, Kathie L., Clement, Simon J., Bazylinski, Dennis A., Kirschvink, Joseph L., McKay, David S., Wentworth, Susan J., Vali, Hojatollah, Gibson, Everett K., McKay, Mary Fae, & Romanek, Christopher, S. (2001). Truncated hexa-octahedral magnetite crystals in ALH84001: Persumptive biosignatures, *Proceedings of the National Academy of Sciences*, **98**, 2164–9.

Thompson, Katherine H., & Orvig, Chris. (2003). Boon and Bane of metal ions in medicine. *Science*, **300**, 936–9.

Tilton, G. R. (1988). Age of the solar system. In *Meteorites and the Early Solar System*, J. F. Kerridge and M. S. Mathews, eds. (pp. 259–75). Tucson: University of Arizona Press.

TIME. (2003). *The Secret of Life*. Nancy Gibbs, *A Twist of Fate*. Michael Lemonick.

Tononi, G., Sporns, O., & Edelman, G. M. (1999). Measures of degeneracy and redundancy in biological networks. *Proceedings of the National Academy of Sciences*, **96**, 3257–62.

Troland, Leonard Thompson. (1914). The chemical origin and regulation of life. *The Monist*, **24**, 92–133.

Turing, A. M. (1936). On computable numbers, with an application to the *Entscheidungs Problem. Proc. Lond. Math. Soc. Ser. 2*, **42**, 230–265. [A correction, **43**, 544–6.]

Unsigned. (1970). Central dogma reversed. *Nature*, **226**, 1198–9.

Unsigned. (2002). Wag the dogma. *Nature Genetics*, **30**, 343–4.

Urey, H. G. (1952). On the early chemical history of the Earth and the origin of life, *Proceedings of the National Academy of Sciences*, **38**, 351–63.

Urey, H. G., Dawsey, L. H., & Rice, F. O. (1929). The absorption spectrum and decomposition of hydrogen peroxide by light. *Journal of the American Chemical Society*, **51**, 1371–83.

Urey, H. G., & Lavin, G. I. (1929). Some reactions of atomic hydrogen. *Journal of the American Chemical Society*, **50**, 3286–93.

Urey, Harold C., & Smallwood, Hugh M. (1928). An attempt to prepare traiomic hydrogen. *Journal of the American Chemical Society*, **50**, 620–6.

Uy, R., & Wold, F. (1977). Postranslational covalent modification of protein. *Science*, **198**, 890–6.

Van Zuilen, Mark A., Lepland, Aivo, & Arrhenius, Gustaf. (2002). Reassessing the evidence for the earliest traces of life. *Nature*, **418**, 627–30.

von Neumann, J. (1966). *Theory of Self-Replicating Automata*. Urbana: University of Illinois Press.

von Neumann, John. (1932). *Mathematische Grundlagen der Quantum Mechanic*. Berlin: Springer Verlag.

Wächtershäuser, Günter. (1997). The origin of life and its methodological challenge. *Journal of Theoretical Biology*, **187**, 483–94.

Wächtershäuser, Günter. (1988a). An all-purine precursor of nucleic acids. *Proceedings of the National Academy of Sciences*, **85**, 1134–5.

Wächtershäuser, Günter. (1988b). Pyrite formation, the first enegy source for life: A hypothesis. *Systems and Applied Microbiology*, **10**, 307–4.

Wächtershäuser, Günter. (1988c). Before enzymes and templates: Theory of surface metabollism. *Microbiological Reviews*, **52**, 452–84.

Wächtershäuser, Günter. (1990). Evolution of the first metabolic cycles. *Proceedings of the National Academy of Sciences*, **87**, 200–4.

Wächtershäuser, Günter. (1994). Life in a ligand sphere. *Proceedings of the National Academy of Sciences*, **91**, 4283–7.

Wächtershäuser, Günter. (1998). Origin of Life in an Iron-sulfur world. In *The Molecular Origins of Life*, A. Brack, ed. Cambridge, UK: Cambridge University Press.

Wächtershäuser, Günter. (2000). Life as we don't know it. *Science*, **289**, 1307–8.

Walker, J. C. G. (1976). Implications for atmospheric evolution of the inhomogeneous accretion model of the origin of the Earth. In *The Early History of the Earth*, B. F. Windley, ed. London: John Wiley.

Wang, Lei, Brock, Ansgar, Herberich, Brad, & Schultz, Peter G. (2001). Expanding the genetic code of Escherichia coli. *Science*, **292**, 498–500.

Watson, J. D., & Crick, F. H. C. (1953a). Molecular structure of nucleic acids. *Nature*, **171**, 737–8.

Watson J. D. & Crick F. H. C. (1953b). Genetical implications of the structure of deoxyribonucleic acid. *Nature*, **171**, 964–7.

Watson, J. D. (2003). *DNA: The Secret of Life*. New York: Alfred A. Knopf.

Watson, J. D., Hopkins, N. H., Roberts, J. W., Steiktz, J. A., & Weiner, A. M., eds. (1987). *Molecular Biology of the Gene*. forth edition. Menlo Park, CA: Benjamin/Cummings.

Watson, James D. (1968). *The Double Helix*. New York: A Mentor Book, New American Library.

Watson, James D. (2001). *Genes, Girls, and Gamow*. New York: Random House.

Weber, Arthur L., & Miller, Stanley L. (1981). Reasons for the occurrence of twenty coded protein amino acids. *Journal of Molecular Evolution*, **17**, 273–84.

Webster, Paul. (2003). Prestigious plant institute in jeopardy. *Science*, **299**, 641.

Weiss, O., Jiménez, M., & Hertzel, H. (2000). Information content of protein sequences. *Journal of Theoretical Biology*, **206**, 379–86.

Welton, M. G. E., & Pelc, S. R. (1966). Specificity of the sterochemical relationship between ribonucleic acid-triplets and amino acids. *Nature*, **209**, 870–2.

Wiechert, A. N., Halliday, A. N., Lee, D.-C., Synder, G. A. Taylor, L. A., & Rumble, D. (2001). Oxygen isotopes and the Moon-Forming Giant Impact. *Science*, **294**, 345–348.

Wiener, Norbert. (1948). *Cybernetics*. New York: The Technology Press, John Wiley.

Wilde, Simon A., Valley, John W., Peck, William H., & Graham, Colin M. (2001). Evidence from detrital zircons for the existance of continental crust and oceans on Earth 4,4 Gyr ago. *Nature*, **409**, 175–8.

Wildman, Derek E., Uddin, Monica, Liu, Guozhen, Grosman, Lawrence I., & Goodman, Morris. (2003). Implications of natural selection in shaping 9.4% nonsynonymous DNA identity between humans and chimpanzees: Enlarging genus, *Homo. Proceedings of the National Academy of Sciences*, **100**, 7181–7188.

Wilkins, M. H. F., Stokes, A. R., & Wilson, H. R. (1953). Molecular Structure of Deoxpentose nucleic acids. *Nature*, **171**, 738–40.

Williams, R. J. P. (1990). Iron and the origin of life. *Nature*, **343**, 213–14.

Williams, T., & Fried, M. (1986). A mouse locus at which transcription from both DNA strands produces mRNAs complementary at their $3'$ ends. *Nature*, **322**, 275–8.

Wills, Christopher, & Bada, Jeffrey. (2000). *The Spark of Life*. Cambridge, MA: Peraues Publishing.

Wills, P. R. (1986). Scrapie, ribosomal proteins and biological information. *Journal of Theoretical Biology*, **122**, 157–78.

Wills, P. R. (1989). Genetic information and the determination of functional organization in biological systems. *Systems Research*, **6**, 219–26.

Willson, S. J. (1998). Measuring inconsistency in phylogenic trees. *Journal Theoretical Biology*, **190**, 15–36.

Woese, C. (1998). The universal ancestor. *Proceedings of the National Academy of Sciences*, **95**, 6854–959.

Woese, C. R. (1967). *The Genetic Code.* New York: Harper & Row.
Woese, Carl. (2000). Interpreting the universal phylogenetic tree. *Proceedings of the National Academy of Sciences,* **97**, 8392–6.
Woese, Carl R. (2002). On the evolution of cells, *Proceedings of the National Academy of Sciences,* **99**, 8742–7.
Woese, Carl, Olsen, Gary J. Ibba, Michael & Söll, Dieter. (2000). Aminoacyl-tRNA Synthetases, the genetic code, and the evolutionary process. *Microbiology and Molecular Biology Reviews,* **64**, 202–36.
Wöhler, Friedrich. (1828). Über künstliche Bildung des Harnstoffs. *Poggendorf's Annalen,* **12**, 253–6.
Wolfram, Stephen. (2002) *A New Kind of Science.* Champaign. IL: Wolfram Media Inc.
Wolosker, Herman, Blackshaw, Seth, & Synder, Solomon H. (1999). Serine racemase: A glial enzyme synthesizing D-serine to regulate glutamate-N-methyl-D-aspartate neurotransmission. *Proceedings of the National Academy of Sciences,* **96**, 13409–14.
Wolynes, P. G. (1998). Computational biomolecular science. *Proceedings of the National Academy of Sciences,* **95**, 5848.
Won, Hyosig, & Renner, Susanne S. (2003). Horizontal gene transfer from flowering plants to Gnetum. *Proceedings of the National Academy of Sciences,* **100**, 10824–9.
Wong, J. T.-F. (1976). The evolution of the genetic code. *Proceedings of the National Academy of Sciences,* **72**, 1909–12.
Wong, J. T.-F. (1983). Membership mutation of the genetic code: Loss of fitness by tryptophan. *Proceedings of the National Academy of Sciences,* **80**, 6303–6.
Wong, J. T.-F., & Hong, Xue. (2002). Self-perfecting evolution of heteropolymer building blocks and sequences as the basis of life. In *Fundamentals of life.* Paris: Elsevier.
Wong, J. T.-F. (1975). A co-evolution theory of the genetic code. *Preceeding of the National Academy of Sciences,* **72**, 1909–12.
Wong, J. T.-F. (1981). Co-evolution of the genetic code and amino acid biosynthesis. *Trends in Biochemical Science,* **6**, February, 33–5.
Wong, J. T.-F. (1988). Evolution of the genetic code. *Microbiological Sciences,* **5**, 174–80.
Wu, C.-I., Li, W.-H., Shen, J. J., Scarpulla, R. C. Limbach, K. J., & Wu, R. (1986). Evolution of cytochrome c genes and pseudogenes. *Journal of Molecular Evolution,* **23**, 61–75.
Yamao, F., Muto, A., Kawauchi, Y., Iwami, M. Iwagami, S., Azumi, Y., & Osawa, S. (1985). UGA is read as trytoptophan in *Mycoplasma capricolum. Proceedings of the National Academy of Sciences,* **82**, 2306–9.
Yin, Qinghu, Jacobsen, Yamashita, K. Blichert-Toft. Télouk, P., & Albarède, F. (2002). A short timescale for terrestrial planet formation from Hf-W chronometry of meteorites. *Nature,* **418**, 949–52.
Yockey, Hubert P. (1958). A study of aging, thermal killing and radiation damage by information theory. In *Symposium on Information Theory in Biology.* Hubert P. Yockey, Robert Platzman, & Henry Quastler, eds. (pp. 297–316). New York: Pergamon Press.
Yockey, Hubert P. (1974). An application of information theory to the Central Dogma and the sequence hypothesis. *Journal of Theoretical Biology,* **46**, 369–406.
Yockey, Hubert P. (1977a). A prescription which predicts functionally equivalent residues at given sites in protein sequences. *Journal of Theoretical Biology,* **67**, 337–43.

Yockey, Hubert P. (1977b). On the information content of cytochrome c. *Journal of Theoretical Biology*, **67**, 345–76.

Yockey, Hubert P. (1977c). A calculation of the probability of spontaneous biogenesis by information theory. *Journal of Theoretical Biology*, **67**, 377–98.

Yockey, Hubert P. (1978). Can the Central Dogma be derived from information theory? *Journal of Theoretical Biology*, **74**, 149–52.

Yockey, Hubert P. (1979). Do overlapping genes violate molecular biology and the theory of evolution? *Journal of Theoretical Biology*, **80**, 21–26.

Yockey, Hubert P. (1981). Self-organization origin of life scenarios and information theory. *Journal of Theoretical Biology*, **91**, 13–31.

Yockey, Hubert P. (1990). When is random random? *Nature*, **344**, 823.

Yockey, Hubert P. (1992). *Information Theory and Molecular Biology*. Cambridge, UK: Cambridge University Press.

Yockey, Hubert P. (1995a). Information in bits and bytes. *BioEssays*, **17**, 85–7.

Yockey, Hubert P. (1995b). Comments on "Let there be life: Thermodynamic reflections on biogenesis and evolution" by Avshalom C. Elizur. *Journal of Theoretical Biology*, **176**, 349–55.

Yockey, Hubert P. (1996). Second challenge to Oparin Medal. *ISSOL Newsletter*, **23**, 34.

Yockey, Hubert P. (1997). Walther Löb, Stanley L. Miller, and prebiotic "building blocks" in the silent electrical discharge. *Perspectives in Biology and Medicine*, **41**, 125–31.

Yockey, Hubert P. (1997). Haeckel, C. Darwin, A. Oparin, Stanley Miller, Walther Löb and the History of the Origin of Life. Invited paper at *The Chemistry of Life's Origins*, part of the Northeast Regional Meeting of the American Chemical Society, June 23–24.

Yockey, Hubert P. (1997d). Life on Mars? Did it come from Earth? *Origins and Design*, **18**, 10–15.

Yockey, Hubert P. (1998). Life on Mars from ALH84001? Revisited, *Origins and Design* **19**, 4–5.

Yockey, Hubert P. (1998). Life on Mars From ALH84001? revisited. *Origins and Design*, **19**, 4–5.

Yockey, Hubert P. (2000). Origin of life on Earth and Shannon's theory of communication. In Open problems of computational molecular biology. *Computers and Chemistry*, **24**, 1, 105–23.

Yockey, Hubert P. (2001). Behe's irreducible complexity and evolutionary theory. *Reports*, **21**, 18–20.

Yockey, Hubert P. (2002a). Information theory, evolution and the origin of life. In *Fundamentals of Life*, Gyuala Pályi & Luciano Cagliotti, eds. Paris: Elsevier.

Yockey, Hubert P. (2002b). More light on pioneers of electrochemistry. *Nature*, **415**, 833.

Yockey, Hubert P. (2002c). Information theory, evolution and the origin of life. *Information Sciences*, **141**, 219–25.

Yockey, Hubert P. (2003). Comment on "Some like it hot, but not the first biomolecules." *Science*, **296**, 1982–3.

Yockey, Hubert P., Platzman, Robert P., & Quastler, Henry, eds. (1958a). *Symposium on Information Theory in Biology*. New York: Pergamon.

Zaug, A., & Cech, T. R. (1986). The intervening sequence of RNA of *Tetrahymena* is an enzyme. *Science*, **231**, 470–5.

Zinoni, F., Birkman, A., Stadtman, T., & Böck, A. (1986). Nucleotide sequence and expression of the selenocysteine-containing polypeptide of formate dehydrogenase (formate-hydrogen-lyase-linked) from Escherichia coli. *Proceedings of the National Academy of Sciences*, **83**, 4650–4.

Zinoni, F., Birkman, A., Stadtman, T., & Böck, A. (1987). Co-translational insertion of selenocysteine into formate dehydrogenase from Escherichia coli directed by a **UGA** codon, *Proceedings of the National Academy of Sciences*, **84**, 3156–60.

Zinoni, F., Heider, J., & Böck, A. (1990). Features of the formate dehydrogenase mRNA necessary for decoding of the **UGA** codon as selenocysteine. *Proceedings of the National Academy of Sciences*, **87**, 4660–4.

Zuev, Peter S., Sheridan, Robert S., Albu, Titus V., Truhlar, Donald G., Hrovat, David A., & Borden, Westin Thatcher. (2003). Carbon tunneling from a single quantum state. *Science*, **299**, 867–70.

Zurek, W. H. (1984). Reversibility and stability of computational systems. *Physical Review Letters*, **53**, 391–4.

Zurek, W. H. (1989). Thermodynamic cost of computation, algorithmic complexity, and the information metric. *Nature*, **341**, 119–24.

Index

aaRSs (aminoacyl-tRNA synthetases), 115
abstract euclidean vector space, 59
  amino acids in, 59–68
  BH distance of separation, 65
ad hoc scores, 48
Ala-Tyr pairs
  Hamming distances in, 63
  in iso-1-cytochrome c, 68
algorithmic information theory, 168, 170
Alice in Wonderland (Carroll), 164
Alzheimer's disease, 57
American Standard Code for Information
    Interchange. *See* ASCII
amino acids
  in abstract euclidean vector space, 59–68
  cyclic imino acids versus, 83
  enclosing spheres for, 61
  formation of, 126–127
  functionally equivalent, 57, 60, 77–80,
    88
  Gly-Ser mutations in, 81
  Hamming distances within, 76, 82
  in iso-1-cytochrome c, 68
  non-lethal codons, 94
  peptide bond formation, 142–143
  polymerization of, 130–131
  primitive coding templates for, 130
  probabilities, 55
  protein synthesis for, 110
  protein-folding pathways for, 57
  supernumerary, 110
  thyroid gland deficiency and, 57
  Thyroxine, 57
aminoacyl-tRNA synthetases. *See* aaRSs
Ansatz, 59. *See also* prescriptions

α1-antitrypsin
  in iso-1-cytochrome c, 84
Aquinas, Thomas, 176
Aristotle, 181
Arrhenius, Svante, 142
Arrowsmith (Lewis), 121
ASCII (American Standard Code for
    Information Interchange), 14, 17, 19,
    117
  alphabet members of, 17
  bytes in, 17
Astrobiology (journal), 142
Atlas of Protein Sequence and Structure 5,
    57
autotrophs
  in citric acid cycle, 146
  CO_2 in, 146
Avery, 8

Bacillus brevis, 118
Bacillus subtilis, 110
Bada, Jeffrey, 116, 119, 123, 126–127, 146,
    148
Baltimore, David, 22
banded iron formation. *See* BIF
Baudisch, Oskar, 124, 126, 148
Bayes Theorem, 49
Beagle 2, 141
Behe, Michael, 178, 181
Bergson, Henri, 150
BIF (banded iron formation)
  on Earth, 138
Big Bang theory, 13
  First Cause within, 133–135
  origin of matter and, 133

binary alphabets, 15, 97
  bits in, 15
  in electronic communications, 97
  isomorphic, 21
  members of, 16
  mutual entropy and, 47
  proteome, 21
  sense code letters, 16
biosynthetic pathways
  for codons, 107
bit strings, 170
bits
  in binary alphabets, 15
  bytes versus, 15
blending inheritance theory, 3
block codes, 16
  error detection properties of, 39–40
Bohr, Niels, 5, 175
Boltzmann constant, 25, 31
Bovine pancreatic trypsin inhibitor, 82
Brenner, Sydney, 13
Bryan, William Jennings, 178
Bungenberg de Jong, H.G., 146, 157,
  188
bytes
  in ASCII, 17
  bits versus, 15
  in United States Postal Code, 16

Calvin, Melvin, 13
carbon
  compound composition for, 138
  fixation, 132
  in "origin of life" events, 131
  yields, 127
Carreras, José, 4
Carroll, Lewis, 7, 164
Caruso, Enrico, 4
The Central Dogma, 20, 21, 23, 52, 132,
  160
  absolute temperature in, 25
  energy dissipation and, 24
  misconceptions of, 22–24
  "origin of life" and, 147
  restrictions of, 21, 182
  Sequence Hypothesis vs., 23
  tenets of, 20
  three transfers in, 20
Chaitin, Gregory, 2, 168, 169
Channel Capacity Theorem, 4, 42–46, 47, 49,
  158, 162, 181, 184
  communication systems within, 42
  conditional entropy and, 42
chaos theories
  order from, 164–165
Chargaff, Edwin, 13

Chemiker-Zeitung, 124
"chemistry of life," 8
  nucleic acid in, 8
chiral molecules, 2
Chironomus thummi, 36
citric acid cycle, 146
  autotrophs in, 146
Clinton, Bill, 139, 140
clostridial type ferrodoxins
  in Jukes Proposal, 98
Clostridium butyricum, 98
Clostridium thermoaceticum,
  106
code letters
  errors within, 16–17
  Hamming distances and, 16
  non-sense, 16
  sense, 16, 39
  source, 40
Code of Hammurabi, 13, 117
codes
  ASCII, 14
  block, 16, 17
  Code of Hammurabi, 13
  complete, 41
  formal definition of, 14
  Morse, 11, 14
  redundant, 183
  United States Postal Code, 14
  Universal Product Bar Code, 14
  versus "sequences," 10–11, 13–14
"code-script," 4
coding theory, 3, 113
  definition of, 13
  white noise applications, 37
codons
  biosynthetic pathways for, 107
  cylinder, 101, 108
  division of, 53
  in genetic codes, 19, 95
  in Jukes Proposal, 104
  for mitochondria, 94, 103, 112
  in molecular biology, 15
  in mRNA, 15
  non-lethal, 94
  triplet formation, 104–106
  in tRNA, 17
column vectors, 14
communication systems
  DNA-mRNA-proteome, 33–34
  genetic, 33
  isomorphism between, 170
communication theory, 24
complexity
  information theory and, 169
  phylogenetics and, 169

randomness and, 168
Shannon entropy and, 168
conditional entropy, 42, 43–46
Corey, 9, 10
Cramér's Rule, 67
creationism, 176
   Intelligent Designer as part of, 176
Crick, Francis H.C., 3, 8, 9, 10, 13, 17, 20, 21,
   24, 96, 97, 104, 112
   "code" versus "sequence" for, 10–11
   DNA breakthrough for, 3, 8
   sequence hypothesis of, 57
Critique of Judgment (Kant), 150
Curie, Frédéric, 125
Curie, Irène Joliot, 125
cyclic imino acids
   amino acids versus, 83
Cys amino acids, 82
cytosine
   in DNA, 145
   in mRNA, 145

Darwin, Charles, 4, 119–121, 141, 150, 177
   on blended characteristics, 179
   "missing links" for, 4
   "One long argument" of, 4
Darwin, Francis, 119
Darwinian Threshold, 174
Darwinism, 184
   religious clergy and, 177–178
Darwin's Black Box (Behe), 178
Dayhoff matrix, 59
Dayhoff, M.O., 57
"degrees of homology," 49
Delbrück, Max, 13
Descartes, René, 14
   sample spaces for, 14–15
dialectical materialism, 12, 151–152
   *Law of the Transformation of Quantity into
      Quality*, 151
   NASA and, 182
   quantum mechanics and, 12
   Shklovskii, Iosef Samuilovich, and,
      155–156
Dialectics of Nature (Engels), 156
The Dialectics of Nature (Engels), 151
*Die Koazervation und ihre Bedeutung für die
     Biologie* (Bungenberg de Jong), 157
digital electronics, 4
Directed Panspermia theory, 142
discrete memoryless source, 43
DNA
   cell importance of, 9
   Crick, F.H.C., and, 8
   cytosine in, 145
   as hereditary source, 150

"orderly" sequences for, 185
templates, 6
transcription, 26
in Turing machines, 171
Watson, J,D., and, 8
DNA-mRNA-proteome communication
   system, 33–34
   isomorphism in, 33
   Markov process in, 33
   transcription in, 34
   translation during, 34
Dodgson, Charles Lutwidge. *See* Carroll,
   Lewis
Domingo, Plácido, 4
Doty, Paul, 13
doublet codons
   functions for, 100
   in mRNA, 15
Dounce, Alexander, 13
Drake, Richard Frank, 187
Driesch, Hans, 150
Drosophilia melanogaster, 86
Drosphilia yakuba, 105
Duve, Christian de, 119, 187
The Dynamics of Living Matter (Loeb), 121

Earth
   age of, 135
   BIF and, 138
   early life on, 138–139
   planet formation and, 95
   radioactive decay series and, 138
Eck, R.V., 57
Eigen, Manfred, 32, 158
   "error catastrophe" for, 162
   on information theory, 158–159
   "master sequence" proposal by, 30, 159
Einstein, Albert, 173
electricity, 122–133
electro-chemistry, 122–123
electron transfer pathways
   iso-1-cytochrome c in, 59
   prescription in, 59
Elements (Euclid), 172
energy dissipation, 26
   in *The Central Dogma*, 24
   minimum, 25, 26
Engels, Friedrich, 118, 151, 156, 157
   on evolution theory, 151
entropy
   conditional, 42, 43–46
   in information theory, 31
   mutual, 43
   negative, 28, 31–32
   in probability theory, 31
   von Neumann, 31

entropy of probability theory, 14, 185
  entropy in, 31
  sample space in, 14
Entscheidungsproblem (decision problem),
  170
"error catastrophe," 162–163, 184
  Gompertz function within, 162
error detection
  for block codes, 39–40
evolution theory, 107
exobiology, 142–143
"Experiments on plant hybrids" (Mendel), 156

Falwell, Jerry, 178
ferrodoxins, 106
  clostridial type, 98
Feynman, Richard, 13
finite schemes, 14
Fisher, Ronald Aylmer, 93
Five Billion Vodka Bottles to the Moon
  (Shklovskii), 155
Five Proofs (Aquinas), 176
formic acid, 126–127
Fox, Sidney, 135
Franklin, Rosalind, 9, 10

Galileo, Galilei, 183
Gamow, George, 8, 10, 11, 14, 58, 93, 180, 184
  biology contributions of, 12–13
  genetic specificity for, 11
  "number of the beast" for, 10
  quantum mechanical tunnelling applications
    of, 12
  RNA Tie club of, 13
  sequence hypothesis of, 58
Gamow, Rho, 12
*Gatlinburg Symposium on Information Theory
  in Biology*, 11–12
genetic codes, 19, 50
  amino acids in, 110
  codons in, 19, 95
  complete, 42
  degenerate, 15, 42
  error correction effects in, 41–42, 94
  evolution of, 113, 181
  extensions of, 102
  Hamming distances in, 105, 109
  instantaneous factors for, 107–108
  "magic numbers" within, 110–111
  mapping relationships within, 96
  Markov states in, 101, 103, 109
  mitochondrial, 111–112
  for mRNA, 18
  mutations in, 101
  optimal factors for, 108–109

origins of, 93
overlapping genes and, 87–92
permutations in, 95
pons asinorum, 101
redundant, 15, 20–21
source alphabet for, 19
transmission as part of, 35
vocabulary expansion of, 103–104
genetic communication systems
  components of, 33
  errors in, 38
  gene duplication rates in, 38, 181
  majority logic redundancy within, 38
  sense code letters within, 39
  stochastic process in, 33
genetic noise
  definition of, 37–38
  minimization of, 101
  missense within, 37
  reduction effects for, 41
  stochastic Markov process within, 37
  white, 52
genetic systems
  communication, 33, 38, 39
  information, 3, 7, 159–160, 184
  logic, 26
genetics
  longevity interactions, 163
  orderliness in, 167
  transmission of, 4
genomes
  error processes in, 180–181
  in information theory, 6
  tunnelling processes in, 180
Gilbert, William S., 115, 116
Glaucon, 1
glycine
  formation of, 123
Gly-Ser mutations
  in amino acids, 81
Gödel, Kurt, 174, 188
Goldin, Daniel, 139
Gordon, H., 13
Göttingen School, 161
  on *Urschleim*, 161
gramicidin A, 118
"Great chain of being" (Aristotle), 181
Griffith, 17
GULAG Archipelago (Solzhenitsyn), 157
Gulliver's Travels (Swift), 147

Haeckel, Ernst H.P.A., 114, 115, 126, 131,
  156, 157, 186
  on "origin of life," 115–116
  on Pasteur, Louis, 115

on randomness, 161
on *Urschleim*, 117, 146–147, 160, 182
Haldane, John B.S., 115, 126, 146, 148, 175
  contributions of, 127–128
  on "origin of life" events, 127–128
Hamming distances, 40–41
  for Ala-Tyr pairs, 63
  in amino acids, 76
  code letters and, 16
  in genetic codes, 105, 109
  in iso-1-cytochrome c, 84
  in Jukes Proposal, 98, 99
  synonymous source code letters and, 40
  white noise and, 41
Hamming, Richard, 16
Harpers Magazine, 22
Hegel, Georg W.H., 149
Helmholtz, Hermann Ludwig Ferdinand von, 142
Hertzsprung–Russell diagram, 133
Hesiod (philosopher), 144
Hilbert, David, 170, 171
homochirality, 118
homologous protein families, 47–48
  similar versus, 47
Hooker, Joseph, 119
horizontal gene transfer, 174, 181
  cell location for, 174
Horowitz, Vladimir, 170
Human Genome Project, 23, 186
Hume, David, 176
Huxley, Thomas Henry, 177

Incompleteness Theorem, 174
  Turing machines and, 171
information theory, 3, 6, 31, 158, 184
  complexity in, 169
  content as part of, 56
  entropy in, 31
  genome as part of, 6
  instantaneous genetic codes for, 108
  sequence symbols in, 6
  Shannon entropy and, 34
  transmission as part of, 34
inheritance
  blending theory, 3
  segregated, 3
Intelligent Designer, 178, 181
  creationism and, 176
Intelligent Life in the Universe
  (Sagan/Shklovskii), 118, 155
"irreducible complexity," 178
iso-1-cytochrome c, 29–30, 53, 62, 65, 68, 83, 87
  Ala-Tyr pairs within, 68

amino acid predictions for, 68
α1-antitrypsin in, 84
BH considerations in, 83
Cys amino acid sites in, 82
in electron transfer pathways, 59, 81
formation of, 58
hamming distances in, 84
history of, 58
information content of, 69–75, 84–85
oligonucleotide mutagenesis in, 82
polypeptide sequences in, 64
R values in, 76
redundancy in, 38
Shannon–McMillan–Breiman theorem and, 84, 118
site-directed mutagenesis in, 82

Johnson, Philip E., 178
Journal of the Chemical Society, 122
Jukes Proposal, 97–101
  clostridial type ferrodoxins in, 98
  codon assignments in, 104
  Hamming distances within, 98
  tests of, 97
Jukes, Thomas, 116
Just So Stories (Kipling), 1, 146

Kant, Immanuel, 150
Kerr, Richard, 187
Kipling, Rudyard, 146
Kolmogorov, 169
Kraft Inequality, 108
Krankenhaus, Rudolf Virchow, 124

*Law of the Transformation of Quantity into
  Quality*, 155
  in dialectical materialism, 151
Lawrence, Ernest Orlando, 125
Laws of Thermodynamics
  First, 31
  Second, 21, 30, 185
Lazcano, 123, 126–127, 146
Ledley, Robert, 13
Lewis, Sinclair, 121
LIFE: Its Nature, Origin and Development
  (Lysenko), 156
"Light and Life" (Bohr), 5
lightning
  corona discharge during, 130
  simulation of, 129–130
Löb, Walther, 122, 123, 124, 126, 127–128, 148
  biometric synthesis apparatus, 137
  life of, 124–125

Loeb, Jacques, 157
  on "origin of life," 121–122
Logic and the understanding of nature
    (Hilbert), 172
Lysenko, Trofim, 153
  Stalin, Josef, and, 154

majority logic redundance
  in genetic communication system, 38
Markov process
  in DNA-mRNA-proteome communication
    system, 33
  stationary, 43, 101
Markov states
  in genetic codes, 101, 103, 109, 181
Mars (planet)
  carbonate minerals on, 140
  exploration on, 141
  life on, 139–141
Mars Express, 141
Marx, Karl, 152, 157
"master sequence" proposal, 30, 159
The Mathematical Appendix, 53
Maxwell–Boltzmann–Gibbs entropy, 26, 32,
    185
Mayr, Ernst, 5, 183
McKay, Christopher, 149
McMillan, 43
mechanist–reductionism, 5, 150–151
  in Soviet Union, 152
Medvedev, Zhores, 156
Mendel, Gregor, 3, 156
  on segregated inheritance, 3
Methanosarcina barkeri, 17
Metropolis, Nicholas, 13
Michurinism–Lysenkoism, 156
The Mikado, 114, 116
Miller, Stanley, 124, 125, 126, 129, 131, 186
  on amino acid formation, 126–127
  contributions of, 125–128
Miller–Urey experiments, 118, 128, 136
"missing links," 4
  for Darwin, Charles, 4
mitochondria
  codons for, 94, 112
  endosymbiotic theory of, 111
  genetic codes within, 111–112
  mRNA editing for, 106–107
molecular biology
  codons in, 15
moons
  planet formation and, 135–138
  Roche limits and, 135
  surface metamorphism for, 137
Morgano–Weismannite genetics, 154–156

Morse Code, 14
  evolution of, 11
Morse, Samuel F.B., 11
mRNA, 34
  alphabet, 21, 36
  cytosine in, 145
  doublet codons, 15
  genetic code for, 18
  mitochondrial codons and, 106–107
  templates, 6
  in Turing machines, 181
Muller, Herman, 153, 156
mutual entropy, 43, 45, 48–49, 92
  binary alphabet similarity within, 47
  definition of, 46
  as information content measure, 53–56
  properties, 46–47
Mycoplasma capricolum, 112
Mycoplasma caprolium, 104, 111
Mycoplasma genitatilium, 85, 174
Mycoplasma pneumoniae, 85

NASA (National Aeronautic and Space
    Administration), 142
  "origin of life" policies of, 157, 182–183
National Aeronautic and Space
    Administration. *See* NASA
Nature (magazine), 21, 85, 139, 148
Naturphilosophie, 149, 151
negative entropy, 28, 31–32
  Shannon entropy theorem and, 32,
    185
negentropy. *See* negative entropy
Nelson Bay Retrovirus, 85
Neurospora crassa, 104
Newsweek, 140
Newton, Isaac, 173
noise, 33
non-sense code letters, 16, 37, 39
nucleotide sequences, 7
Numerov, Boris, 156

Ohno, Susumu, 38
Oken, L., 149
"One long argument," 4
Oparin, Alexandr Ivanovich, 115, 126, 146,
    152, 153, 155, 156
  criticism of, 156–157
  Western True Believers and, 154
Oppenheimer, J. Robert, 188
Opportunity, 141
optical isomers, 2
orderliness
  for DNA, 185
  in genetics, 167

randomness versus, 166
self-organization within, 168
organic formations, 124
Orgel, Leslie, 13, 17
Origin of Life (Haldane), 128
"origin of life" events, 103
carbon compounds and, 138
carbon fixation and, 131
The Central Dogma and, 147
evolutionary process during, 103
Haeckel, Ernst H.P.A., and, 115–116
Haldane, John B.S., and, 128
hypercycles and, 160–161
locations for, 135
Wächtershäuser on, 131–132
origin of matter
Big Bang theory and, 133
*Origin of Species by Means of Natural
Selection or the Preservation of Favored
Races in the Struggle for Life* (Darwin),
120, 177, 179, 182
overlapping genes, 85–86, 87
discovery of, 85
genetic codes and, 87–92
"ozonizer," 122

Paley, William, 176, 177
Paramecium, 104, 111
parity checks, 39
accuracy of, 39
Parkinson's disease, 57
Pasteur, Louis, 114, 115, 117
on optical isomers, 2
scientific discoveries of, 2
Pauling, Linus, 9, 10
Pavarotti, Luciano, 4
percent identities, 48
Perron-Frobenius Theorem, 53, 56
phase space, 31
phylogenetics
complexity and, 169
Darwinian Threshold, 174
universal tree of, 174
Planck, Max, 3
planet formation
Earth, 95
moons and, 135–138
suns and, 133–135
plant life processes, 122
Platonic solids, 172
Platzman, Robert, 11
Poe, Edgar Allen, 2
polypeptide chains, 28
pons asinorum
in genetic codes, 101

Pope, Alexander, 158
Popper, Karl, 131
Pravda, 154
prebiotic soup theory, 97, 114,
182
pyrimidines and, 97
prescriptions, 59
prions, 24
probability amplitudes, 173
protein families
composition of, 116
prebiotic elements of, 123
in Turing machines, 171
protein-folding
disorders, 57
protein-protein recognition, 24
The Protein Information Resource, 65, 68,
76
proteome alphabet, 21
proteomes, 14
protoplasmal primordial atomic globules, 115,
122, 147, 157, 184, 187
The Purloined Letter (Poe), 2
pyrimidines, 97
prebiotic soup in, 97
Pythagorean Theorem, 64, 65

quantum mechanical tunnelling, 12
quantum mechanics, 173
dialectical materialism and, 12
"quasi-species," 159
Quastler, Henry, 11

randomness, 161, 181
complexity and, 168
number generation and, 166
orderliness versus, 166
transcendental numbers and, 167
Redi, Francesco, 114
The Republic (Plato), 1
reverse transcription, 21, 22, 183
reversible computation, 22
ribozymes
RNA and, 144
Rich, Alexander, 13
RNA, 144–145
ribozymes and, 144
in *Urschleim* (primeval slime), 145
RNA Tie club, 13
members of, 13
The RNA World, 144
The Road Not Taken (Frost), 27
Roche limits, 135
Rorshach, Hermann, 1
row vectors, 14

Saccharomyces cerevisiare, 38
Sagan, Carl, 117, 155
sample spaces
    Cartesian, 14
    definition of, 14
    points in, 14
    in probability theory, 14
Schelling, F.W.A., 149
Schopf, J. William, 119, 157
Schrödinger, 4, 32
Science (magazine), 85, 148, 187
scientific theory
    for biology, 2
    for physics, 2
    Socrates and, 1
Scopes Trial, 178
Search for extraterrestrial intelligence. *See*
    SETI
segregated inheritance, 3
selection theory
    chance's role in, 116–117
selenocysteine, 17
sense code letters, 16, 39
sequence hypothesis, 57–59
    The Central Dogma versus, 23
    information content, 59
SETI (search for extraterrestrial intelligence),
    187
Shannon, 28, 43, 45, 158
Shannon entropy theorem, 21, 22, 24, 28, 30,
    32, 43, 52, 159, 160
    complexity as part of, 168
    information flow within, 34
    negative entropy and, 32
    sequence numbers in, 160
Shannon–McMillan–Breiman Theorem, 8, 29,
    30, 59, 85, 118, 158, 185
    definition of, 29
    iso-1-cytochrome sequences in, 84, 118
Shklovskii, Iosef Samuilovich, 117, 155,
    156
    dialectical materialism and, 155–156
Simons, N., 13
Simpson, George Gaylord, 143, 186
    on exobiology, 143
Smith, Maynard, 17
Socrates, 1, 30
Solzhenitsyn, Aleksandr, 157
Sophocles, 3
source code letters, 40
    Hamming distances and, 40
    synonymous, 40
Spallanzani, Lazzaro, 114
spontaneous generation, 114
Spirit, 141

Stalin, Josef, 12, 153
    Lysenko, Trofim, and, 154
Steffens, Lincoln, 154
Stent, Gunther, 13
Stewart, Potter, 2
stochastic process, 37
    in genetic communication system, 33
    markov, 37
    white noise within, 37
Stop Codon Takeover Model, 105, 107
"superiority parameters," 159
supernumerary amino acids, 110
Swift, Jonathan, 147

Teller, Edward, 13
Temin, Howard, 22
templates, 6
    DNA, 6
    mRNA, 6
Tetrahymna, 104, 111
Thermal Emission Spectrometer, 140
Thompson, William, 142
"Through the Looking Glass" (Carroll), 6–7
Thyroxine, 57
transcendental numbers
    randomness and, 167
tRNA, 17
    codons in, 17
Troland, 149
Turing, Alan, 170, 172
Turing machines, 170, 171, 179
    DNA in, 171
    Godel's Incompleteness Theorem and, 171
    logic of, 171
    mRNA in, 181
    protein families in, 171
Twain, Mark, 118

unconstrained channels, 43
United States Postal Code, 14, 15, 16, 17, 19
    bytes in, 16
Universal Product Bar Code, 14, 15, 17, 19
Universe, Life, and Mind (Shklovskii), 118,
    155
unknowables
    for evolution/gene transfer, 173–174
    for genetic code origins, 173
    in mathematics, 171–173
urea
    production of, 150
Urey, Harold, 126, 186
    contributions of, 125–128
Urschleim (primeval slime), 114, 116, 117,
    118, 129–130, 131, 144
    on Earth, 139, 146

Göttingen School and, 161
Haeckel, Ernst H.P.A., and, 117, 146–147,
    160, 182
paradigm for, 132
RNA in, 145
sea foam and, 143

"value parameters," 159
Vavilov, Nikolai Ivanovich, 153, 156
vectors
    column, 14
    row, 14
Versuche über Pflanzen-Hybriden
    (Experiments on plant hybrids) (Mendel),
    179
Virchow, Rudolf, 151
vitalism, 149–150
von Neumann entropy, 31

Wächtershäuser
    on "origin of life events," 131–132
Watson, J.D., 3, 8, 9, 10, 13
    "code" versus sequence for, 10–11

DNA breakthrough for, 8
sequence hypothesis of, 57
"Watson–Crick Theory" (of inheritance), 22
Weiner, Norbert, 31
Western True Believers, 154
white noise
    coding theory applications for, 37
    definition of, 37
    genetic, 47
    Hamming distances and, 41
    stochastic process and, 37
Wilberforce, Samuel, 177
Wilhelm II (Kaiser), 124
Wilkins, M.H.F., 9
Williams, R., 13
Wills, Christopher, 119
Wöhler, Friedrich, 150
The Worker's Paradise, 12

Ycas, Martynas, 11, 13

Zeitschrift für Elektro-Chemie (Journal for
    Electro-Chemistry), 122